OBJECT-BEHAVIOR MODEL
(State Net)

State

Transition

Initial Transition / Initial State

Final State / Final Transition

State-Transition

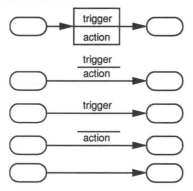

Return to Prior State

Remain in Prior State

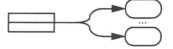

Multiple Prior States Required

Entry into Multiple Subsequent States

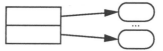

Choice of Prior State

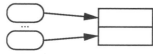

Choice of Subsequent State

State Exception

State Real-Time Constraint

{ rtc }

Path Real-Time Constraint

{ pm1 } ... { pm2 }

{ pm1 - pm2 rtc }

Transition Exception

Transition Real-Time Constraints

{ rtc }
{ rtc }
{ rtc }

rtc: **real-time constraint**

pm: **path marker**

Object-Oriented
Systems Analysis

Object-Oriented Systems Analysis

A Model-Driven Approach

David W. Embley
Barry D. Kurtz
Scott N. Woodfield

YOURDON PRESS
Prentice Hall Building
Englewood Cliffs, New Jersey 07632

Library of Congress Cataloging-in-Publication Data

Embley, David W.
    Object-oriented systems analysis : a model-driven approach / David
W. Embley, Barry D. Kurtz, Scott N. Woodfield.
        p.   cm.
    Includes bibliographical references and index.
    ISBN 0-13-629973-3
    1. Object-oriented programming. 2. System analysis.   I. Kurtz,
Barry D. II. Woodfield, Scott N. III. Title.
QA76.64.E43 1992
004.2'1--dc20   91-26645
                                                                    CIP
                                                                    AC

Editorial/production supervision
    and interior design: *Mary P. Rottino*
Manufacturing buyer: *Susan Brunke*
Prepress buyer: *Mary E. McCartney*
Acquisitions editor: *Paul W. Becker*

© 1992 by Prentice-Hall, Inc.
A Simon & Schuster Company
Englewood Cliffs, New Jersey 07632

The publisher offers discounts on this book when ordered
in bulk quantities. For more information, write:

Special Sales/Professional Marketing
Prentice-Hall, Inc.
Professional & Technical Reference Division
Englewood Cliffs, New Jersey 07632

Printed in the United States of America
10   9   8   7   6   5   4   3   2   1

ISBN   0-13-629973-3

Prentice-Hall International (UK) Limited, *London*
Prentice-Hall of Australia Pty. Limited, *Sydney*
Prentice-Hall Canada Inc., *Toronto*
Prentice-Hall Hispanoamericana, S.A., *Mexico*
Prentice-Hall of India Private Limited, *New Delhi*
Prentice-Hall of Japan, Inc., *Tokyo*
Simon & Schuster Asia Pte. Ltd., *Singapore*
Editora Prentice-hall do brasil, Ltda., *Rio de Janeiro*

Dedication

*To our wives Ann, Rachel, and Linda and our children
for their patience while we spent many extra evenings and weekends
working on this book.*

Contents

Preface

As software systems grow in complexity and size, the task of systems analysis becomes increasingly difficult and time consuming. The challenge faced by systems analysts is to properly capture, maintain, and understand the large volume of information in today's complex software systems.

This book presents a new approach for performing systems analysis and documenting the results of systems analysis. The approach is based on proven methods of information modeling and behavior specification. These methods have been successfully applied to software problems ranging from traditional data processing systems to scientific real-time systems.

The text both explains and shows how to use powerful object-oriented techniques for creating and maintaining analysis models. Many examples of real-life system problems are given that should be familiar to most audiences. By way of contrast, the last chapter of the book gives a solution to a scientific real-time system problem.

The text is intended for systems analysts, software project managers, programmers, consultants, computer science students, and other computer professionals. It is suitable for use in a university-level course on systems analysis and as a supplement for software engineering courses.

An Object-Oriented Approach

Even though object-oriented techniques have been used for many years, the term "object-oriented" is relatively new and sometimes misunderstood. With this in mind, we have anticipated several questions that might be asked about this book.

Why would I, as a C, Pascal, COBOL, or FORTRAN programmer, be interested in anything object-oriented? The analysis technique presented in this book does not require you to change your programming language. Rather, we present a new method for modeling your system to gain greater insight before design and development. You may choose to stay with a procedure-oriented language for implementation. Regardless of your implementation strategy, you will be able to achieve better results based on an accurate and complete systems analysis.

Do I need to have some background in object-oriented techniques before I can understand this book? No, even though some exposure to object-oriented techniques may help, readers with a wide variety of computer science skill levels should find the subject matter and text understandable.

What does "object-oriented" have to do with analysis and information modeling? The basic underlying concepts that make object-oriented languages and object-oriented design successful may also be used to enhance systems analysis. These concepts include developing highly cohesive but independent object classes, viewing an object not only as having static information but also as being able to act and to be acted upon, and exploiting powerful abstraction concepts such as aggregation, classification, and generalization for describing many important, but often-overlooked relationships among system components.

Does "object-oriented" mean "good"? The term "object-oriented" is beginning to mean "good" just as the term "structured" meant "good" in the late 1970s and early 1980s. Even though no one can guarantee that use of a specific method will result in an error-free implementation, we explain how to use the object-oriented frame of mind to its best advantage when modeling systems. The results are usually more accurate and complete and generally better than the results obtained by traditional modeling and analysis techniques.

In summary, the book shows how an object-oriented approach can dramatically improve an analyst's understanding of a system before design and implementation. This improved understanding is the key for increasing product quality and reducing product development costs.

Content and Organization of the Book

The primary purpose of Chapter 1 is to explain why object-oriented techniques are needed for analyzing today's systems. The chapter first discusses the goals and challenges of systems analysis and the relative success of existing analysis techniques. It then gives an overview of our modeling approach to object-oriented systems analysis and lays the groundwork for the rest of the book.

Chapter 2 introduces our method for capturing information about objects and their relationships. It includes our definition of "object" and how this definition compares with those traditionally associated with object-oriented programming languages.

Chapter 2 also includes a discussion of constraints and how they apply to and define objects in object classes and relationships in relationship sets.

Chapter 3 introduces our behavior-specification model. It explains how to model the behavior of individual objects and how objects respond to dynamically occurring events and conditions. The chapter also shows how to model complex behavior including concurrency, exception handling, and real-time constraints.

Chapter 4 explains our technique for organizing and maintaining large system models. It shows how an analyst can produce several views of a system, either for the special needs of different audiences or for reduction in model complexity through abstraction.

Chapter 5 describes high-level object classes, relationship sets, states, and transitions. The availability of high-level modeling components adds another dimension to modeling objects and their relationships and to specifying behavior.

Chapter 6 explains how we model object interactions. Objects within a system model may interact with each other in several ways. We show how to express these interactions in a flexible, yet powerful model. Interaction descriptions tie information modeling and behavior modeling together into a logically complete, fully interacting system.

Chapter 7 discusses model integration. Using model-integration techniques, several analysts may work together to produce large system models. By careful integration, analysts can ensure that object-relationship diagrams, behavior-modeling diagrams, and interaction diagrams are all consistent.

In Chaps. 2 through 7, we present a running example. The example is business-oriented and understandable to a wide audience. In Chap. 8 we present a more technology-oriented example. Our example in Chap. 8 is scientific and exemplifies some of the real-time specification features of our modeling technique.

Appendix A includes a formal definition of our model. The formalism provides the advanced system designer with a solid foundation for our object-oriented, model-driven approach to systems analysis. In Appendix B we briefly discuss how to use our analysis model as the basis for specification and design, in preparation for system implementation. At the end of each chapter we compare the concepts we present with similar concepts presented by others. We also provide references for further reading. Appendix C contains the citations for these bibliographic references.

As a guide for readers, we point out that the basic OSA modeling concepts are found in Chaps. 2, 3 and 6. These chapters present the three basic submodels of OSA: the Object-Relationship Model in Chap. 2, the Object-Behavior Model in Chap. 3, and the Object-Interaction Model in Chap. 6. Mastery of these three chapters is essential to understanding OSA. Chapters 4, 5 and 7 all address the practical problem of how to deal with large, complex analysis tasks. Chapter 4 presents some easy-to-understand techniques for managing large diagrams. When using OSA for analyzing large systems, the diagram-management techniques in Chap. 4 are essential. Chapter 5 contains more advanced material on model construction. For many problems, an analysis can be successfully completed without the material in Chap. 5. The material in Chap. 6 partially depends on some of the basic ideas for high-level views in Chaps. 4 and 5. A light reading of Chap. 4 will provide the reader with a good foundation for Chap. 6. However, a thorough examination of Chaps. 4 and 5 will make Chap. 6 easier to understand.

Neither Chap. 7 nor Chap. 8 is essential for understanding OSA. Instead of introducing new OSA modeling concepts, these chapters provide the reader with further examples of how OSA may be used for analyzing complex systems.

Acknowledgements

The development of OSA as an industry analysis technique began in earnest following the completion of a related graduate thesis project [Kurtz 1988]. Since then, OSA has undergone many evolutionary changes based on research within Hewlett-Packard and at BYU, on consultations with individuals at US West, and on feedback received from several test sites within Hewlett-Packard [Ho 1990, Kurtz 1989a, Kurtz 1989b, Kurtz 1990].

We thank everyone who has helped make the completion of this text possible. We are especially grateful for the help and encouragement of our colleagues at BYU: Lynn Brough, Robert Browne, Stephen Clyde, Christophe Giraud-Carrier, Gary Hansen, Robert Jackson, Stephen Liddle, Thomas McNiel, Yai-Yin Mok, Eric Mortensen, and Hui-Jun Yan; at Hewlett-Packard: John Burnham, Dennis Barnett, Tim Hickenlooper, Donna Ho, Steve Joseph, Teresa Parry, Jagi Shahani, and Vishwannth Kasaravalli; and at US West: Rudolphe Nassif.

Most of all, we appreciate our families for their support and understanding during the many long days and weekends spent to complete this book.

1

Introduction

1.1 Systems

A *system* is a group of interacting objects. The objects, for example, may be people, documents, machines, and data. The interaction among objects can be understood by examining the relationships among the objects, the behavior of each individual object, and the mutual behavior of cooperating objects.

Most of us participate in or observe systems in daily life. For example, when a bank customer cashes a check, the customer participates in the bank's check processing system. The objects involved in this system include the customer, the check being cashed, the bank teller, and the customer's bank account. Figure 1.1 shows these objects of a check processing system with lines representing some of the interactions between the objects. All objects in the check processing system interact and work together to produce a specific result—cash returned to the customer corresponding to the amount declared on the check.

Objects may themselves be considered systems. For example, a copy machine is a system of smaller objects that interact within the machine to produce a duplicate of a printed document. Understanding the definition and nature of systems is the first step in comprehending systems analysis.

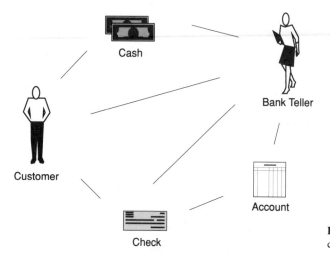

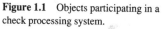

Figure 1.1   Objects participating in a check processing system.

1.2 Systems Analysis

There are many definitions for systems analysis. For some people, systems analysis is the process of modeling a system in its environment. Others refer to systems analysis as defining a problem. Still others refer to it as designing a solution to a problem. We define *systems analysis* as the study of a specific domain of interacting objects for the purpose of understanding and documenting their essential characteristics. The key words in this definition are "study," "understanding," and "documenting."

Although the reason for doing systems analysis need not be to prepare for design and implementation, it often is. When the eventual aim is system implementation, analysis is the process by which we study and understand a system to be implemented and by which we document the results of this understanding.

Understanding a system before design is critical to the success of creating today's complex software systems. For this reason, system designers have searched for the best approach to gain an understanding of a system early in the development process.

1.3 Approaches to Systems Analysis

Many approaches to systems analysis have been tried with varying degrees of success. The approach we advocate in this text is object-oriented. Before immersing ourselves in object-oriented ideas, however, we wish to explain why an object-oriented approach is likely to lead to greater success in understanding and documenting systems than the more traditional natural-language and process-oriented approaches.

1.3.1 Natural Language Analysis

Early systems-analysis documents relied heavily on natural language descriptions. This usually entailed writing page after page of text in an attempt to describe the desired functionality of the system. Since the description of system functionality was of prime importance, the information in natural-language documents was usually organized with respect to system processes. This, along with the linear structure of natural language documents, caused the information about objects and their behavior to be spread throughout the document and made it difficult to map documented concepts back to real-world objects in the system. Figure 1.2 illustrates this difficulty by showing the complex interrelationships between natural language text and system objects.

Natural language may have been adequate for documenting small systems, but, by itself, natural language has proven to be too ambiguous and imprecise for documenting systems of the size and complexity needed today. Computer professionals have therefore searched for other analysis techniques that communicate system characteristics in a clearer, more formal way.

1.3.2 Process-Oriented Analysis

A popular approach to analysis that helps enforce rigor and reduce ambiguity is structured analysis. In this approach analysts describe systems as a network of interacting processes. The approach includes a description of the data used by the processes. These data descriptions are recorded in a data dictionary.

Although the process-oriented approach has occasionally led to success, some problems have been observed. These problems tend to draw analysts prematurely into design. One problem is that the data dictionary entries have more to do with program implementation than with an abstraction of objects in the system domain. Also, the data dictionary is often considered to be secondary in importance, which diminishes its role in assisting analysts to concentrate on analyzing the components in the system. Another problem is that the focus on processes can lead analysts to concentrate on the details of how processing should be done. This approach steers the analyst away from studying system components and their interrelationships towards studying how the system might be designed and implemented. Focusing on "how" leads to a design-oriented study of the system usually before the system is well understood.

Another drawback of structured analysis is the difficulty of mapping concepts between a network of processes and objects existing in a real-world system. There are at least three reasons for this difficulty: (1) Since a process' main connection to a real-world system is through a process name describing a system operation, it is not obvious

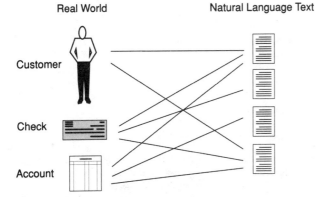

Figure 1.2 Interrelationships between natural language text and real-world objects.

Real World   Analysis Specification

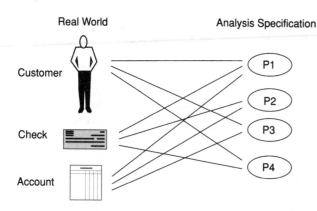

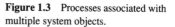

Figure 1.3   Processes associated with multiple system objects.

how the processes relate to real-world objects. (2) As Figure 1.3 shows, most processes associate with multiple system objects. (3) States of system objects and relationships among objects are buried within the process network and scattered among the processing details.

Some data-driven methods try to overcome these problems by organizing operations with respect to objects in the system. While such an organization emphasizes objects, it merely groups operations with respect to objects. Relationships among objects are still missing. Most data-driven methods are really just process-oriented systems organized by objects.

1.3.3 Object-Oriented Analysis

A more recent approach to systems analysis, which shows great promise, organizes all the information in an analysis document around objects in a system. An object-oriented approach modularizes an analysis document along the same object boundaries that exist in a real-world system. Figure 1.4 shows that there is a straightforward correspondence between system objects and analysis documentation components when object-oriented techniques are applied. This is a significant benefit of object-oriented analysis. In addition to simplifying conceptual correspondence, the object-oriented approach also organizes all knowledge about each system object in a single logical location in the analysis

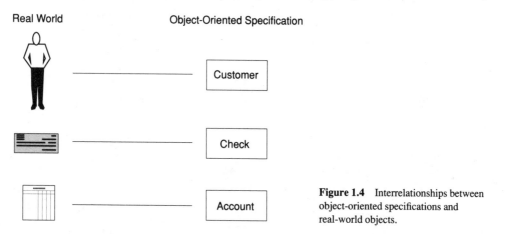

Figure 1.4   Interrelationships between object-oriented specifications and real-world objects.

document. Thus, information about a system object is often easier to locate in an object-oriented analysis than in other analysis approaches.

Because of the emphasis on objects and the corresponding de-emphasis on processes, the object-oriented approach encourages an analyst to concentrate on "what" rather than on "how." The necessary functional properties of a system may easily be expressed, but these do not take priority over the discovery of system objects and their relationships and interactions. This approach reduces the temptation to skip prematurely to design.

Object-oriented techniques also provide forms of abstraction not found in other techniques. In addition to aggregation which is found in many methods, object-oriented techniques provide generalization and classification. The use of these abstractions make it even easier to understand information organized with respect to objects.

The analysis technique we present in this text is object-oriented. We call our analysis method Object-oriented Systems Analysis (OSA). Many benefits of OSA, including those listed, will become evident as we introduce, discuss, and illustrate the various features and concepts of OSA.

1.4 Model-Driven Analysis

In addition to being object-oriented, OSA is also model-driven. Analysts typically use one of two approaches to gather information and document systems. One is a method-driven approach, the other is a model-driven approach.

A method-driven approach consists of a fixed sequence of steps to follow. In practice, the fixed sequence of steps cannot always be followed exactly. When problems are encountered, an experienced analyst adjusts steps and procedures and makes exceptions to rules as may be necessary to complete the task. The successful analyst relies on underlying principles and experience to make appropriate exceptions. Unfortunately, these underlying principles are often undocumented and poorly understood. As a result, inexperienced analysts are usually frustrated when they try to blindly follow the fixed steps of a method-driven approach.

Another approach to analysis concentrates on building a model of the system under study. A model captures specific characteristics exhibited by system objects, and model construction drives the process of acquiring knowledge and asking questions about the system. Instead of a prespecified order of construction, model-driven techniques provide a prespecified set of fundamental concepts with which to model the system under study. Analysts, guided by these fundamental modeling concepts, build models as best suits their needs for gaining insight and understanding.

To illustrate the differences between a method-driven approach and a model-driven approach, we present an analogy. Assume Mary and Susan are seeking employment in a city neither has visited before. Both stay at the same hotel. Both have interviews with the same company. To get to the interview, Mary received instructions provided by a friend. These instructions were supposedly the fastest way to get to the company's office. Mary's method-driven approach is as follows:

> Go four blocks north and then proceed east until you get to the library. Then, turn left and go until you get to the second stop light and turn right. The office is two more blocks on the right.

Unfortunately, Mary misses the library and turns left at a school instead of the library. She gets lost, and must stop and ask someone for help.

Susan's model-driven approach is as follows:

> Susan was told that the city was laid out in a North, South, East, West grid pattern and that each block was numbered in increments of 100 from the center of the city. She also knows that the hotel is at 500 South and 400 East and that the office is at 1650 East and 300 North.

Based on her internal conceptual model of the layout of the city, Susan travels North until she reaches 300 North and then travels twelve and a half blocks East where she finds the office.

Analysts using a model-driven technique can take an approach similar to Susan's. To reach her goal, she first spent time understanding the model of the layout of city streets. Then, using this model she reached her goal quickly and efficiently. Similarly, analysts using a model-driven approach take time to build and understand models directly reflecting the system under study. After that, they know where they are, where they are going, and they know when they reach their desired destination. Most important, they know what to do if they get lost.

Those using a method-driven technique may approach analysis like Mary. She tried to reach her goal by following instructions having no reference to the model of the city layout. Analysts taking this approach may only be successful if they follow the steps precisely and make no mistakes. If they get lost or if anything prevents them from following the steps exactly as specified, they must stop and hope to find an expert for directions.

We prefer the model-driven approach to systems analysis and use this approach in OSA. Our approach does not preclude the development of rules or steps to ease the development of a logical model of reality. The approach does, however, imply that the most important thing to learn is not a step-by-step procedure, but rather the conceptual framework behind the analysis technique.

1.5 Representing Reality

Being object-oriented and model-driven is not enough. An analysis model must also have the expressive power to represent reality easily. If an analyst has to shoehorn ideas into an analysis model in a way that does not seem natural, the model is usually lacking in expressive power. Although system implementation may eventually require shoehorning, the shoehorning should be done during design, not during analysis. An analyst has enough to do just studying, understanding, and documenting a system, without also having to transform ideas into an overly constrained system model.

For example, consider an analysis model that does not allow an object to do more than one activity at a time. Now try to model the high-level behavior of a simple computer printer that loads a new sheet of paper when it is nearly through printing the current sheet. With a single-activity object model, we would have to stop printing to check how much paper is left and then stop longer if the paper must be loaded. Another approach is to contrive other objects such as a print mechanism and a loader mechanism that may not exhibit internal concurrency. But how far do we go? What if the print mechanism exhibits further concurrency? If we try to break down the object of interest too far we get into the business of designing an implementation instead of performing analysis. An analysis model that can express concurrent activities for a single object enables us to avoid premature or unnecessary decomposition of objects during analysis.

Because systems analysis models and systems analysis techniques have historically been developed by people who were first programmers, analysis models usually have been extended programming models with their inherent limitations. Using a programming model to do systems analysis can be like forcing square pegs into round holes. Just as we must shave off the corners of a square peg to make it fit the round hole, we also shave off important concepts to fit our views into restricted analysis models. Using extended programming models of analysis forces us to represent all activities and objects of a system under study in terms of programming-language concepts. Even those objects and actions that will not be implemented require some type of programming. For instance, if we are analyzing an air-traffic-control system, we would be forced to "program" both a pilot and an air-traffic controller even though they will not be part of the final air-traffic-control software.

We have designed our OSA model with the idea in mind that we want to represent reality instead of some particular programming language. For example, we allow objects to be simultaneously involved in more than one activity. Also, we designed all our modeling components to allow an analyst to capture anything of importance about the system. In doing so, we have tried to shed our programming biases. The OSA model encourages an analyst to represent systems the way they are perceived, without any constraint on the way they will be implemented.

1.6 Degree of Formality

Analysts are not mathematicians. We cannot expect an analyst to document understanding in first-order predicate calculus or any other highly formal notation. Analysts should be free to make notes, sketch ideas, and evolve models without the cumbersome task of casting ideas into formal mathematics.

Although formal expressions of system characteristics are not required in OSA, the underlying concepts of OSA are based on formal definitions of system data and behavior modeling. Analysis through the construction of system models whose modeling constructs are based on formal definitions is helpful for several reasons. (1) Models based on formal definitions can provide a foundation for testing model integrity and, to a degree, completeness of an analysis. (2) Since formal model definition ensures a consistent interpretation, it can provide a mechanism for communicating system understanding within the analysis team. (3) For the same reason, a model with a formal foundation can also improve communication to parties outside the analysis team.

A formal definition of the modeling constructs of OSA is in Appendix A. Our presentation in the body of the text is by explanation, discussion, and example.

1.7 An OSA Overview

The remainder of this book describes our OSA model and shows how to use it to do systems analysis. Before describing the details of the model in the remaining chapters, we wish to introduce our model here, by presenting a high-level overview of the central concepts.

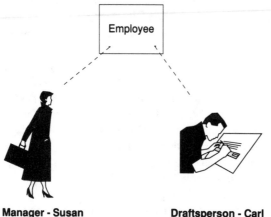

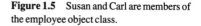

Manager - Susan   **Draftsperson - Carl**

Figure 1.5 Susan and Carl are members of the employee object class.

1.7.1 Object-Relationship Models

The foundation of OSA system models is the object class. An object class represents a group of system objects that have similar characteristics and behavior. In OSA, we denote object classes by rectangles. The rectangle in Figure 1.5 depicts an object class called *Employee*. The labeled rectangle represents a set of objects, each of which is an employee.

Let us assume that a manager named Susan and a draftsperson named Carl are two members of the *Employee* object class. In object-oriented terms, each member object is called an *instance* of the object class. Figure 1.6 shows the commonality of characteristics required of instances of an object class. The circle on the top in Figure 1.6 represents all the characteristics and behavior of manager Susan. The circle on the bottom represents all the characteristics and behavior of draftsperson Carl. The area where the circles overlap includes the common characteristics and behavior that allow each worker to be a member of the *Employee* object class. Some of the common characteristics of employees may be employee number, hire date, and tax information. Common behavior of employees may include receiving a pay check, filling out vacation time

Figure 1.6 Common worker characteristics allowing objects to belong to the employee class.

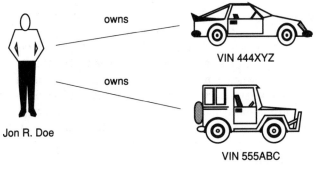

Jon R. Doe

VIN 444XYZ

VIN 555ABC

Figure 1.7   Relationships between a person and two vehicles.

vouchers, and participating in an annual job performance review. The areas of the circles that do not overlap imply that Susan and Carl have characteristics and behavior that differ. Perhaps managers are salaried while draftspersons are hourly workers, and managers make hiring decisions while draftspersons create mechanical drawings.

System objects rarely stand alone. Normally, they have relationships with other system objects. Assume, for example, that we are analyzing a system that keeps track of vehicle ownership. Figure 1.7 shows two relationships that we may find in a real-world system between a person and two vehicles. The relationships say that Jon R. Doe owns the Sports Car with Vehicle Identification Number (VIN) 444XYZ and the Off-Road Vehicle with VIN 555ABC pictured in Fig. 1.7. Our understanding of the real world tells us that the two *owns* relationships have a common meaning. In OSA, we group relationships with a common meaning into a group called a relationship set. Figure 1.8 shows that we represent a relationship set as a line and label the line with a name that denotes the relationship we wish to express.

Diagrams depicting relationship sets, coupled with associated object classes, are the basic components of an OSA Object-Relationship Model. Figure 1.8 is a diagram representing all vehicle owners of interest in the system, all vehicles of interest, and all ownership relationships of interest. An instance of the object class *Vehicle Owner* may be Jon R. Doe. Two instances of object class *Vehicle* may be Jon's Sports Car (VIN 444XYZ) and Jon's Off-Road Vehicle (VIN 555ABC). If so, then two instances of the relationship set "*Vehicle Owner owns Vehicle*" may be the individual relationships "*Jon R. Doe owns Sports Car (VIN 444XYZ)*" and "*Jon R. Doe owns Off-Road Vehicle (VIN 555ABC)*."

Chapter 2 discusses the details of object-relationship models.

1.7.2 Object-Behavior Models

Another part of an analysis document is a description of the behavior of each object in the system. Behavior has three basic components. (1) The states that each object exhibits throughout its existence. (2) The conditions that cause an object to make the

Vehicle Owner
owns Vehicle

Vehicle Owner   Vehicle

Figure 1.8   Figure 1.8   An object-relationship model.

transition from one state to another. (3) The actions performed by, or on, an object in various states and transitions.

Figure 1.9 shows a lamp in two states: off and on. We may cause the lamp to make a transition from the off state to the on state or from the on state to the off state by activating a switch. Activation of the switch is the condition that triggers the transition. In the transition from the off to the on state, the action of applying power to the filament occurs, and in the transition from the on to the off state, the action of removing the power from the filament occurs. Thus, the two states: on and off, the transition condition: switch activation, and the actions: applying power to or removing power from the filament, describe the behavior of a lamp.

In OSA, an analyst may document states, transition conditions, and actions in object-behavior models, which we call state nets. A state net describes the common behavior of objects in an object class. Since each object class has its own state net, OSA naturally groups the behavior of objects with the object class. This aids modularization and the traceability of objects in a system to their corresponding OSA documentation.

Figure 1.10 shows an OSA state net that abstractly models the behavior of a particular lamp for the object class *Lamp*. The rounded rectangles in the diagram with labels *off* and *on* represent states. The rectangular boxes with horizontal dividing lines represent transitions that may take place based on the current state of a lamp. The arrows in and out of a transition give the direction of the transition.

Above the horizontal line in the transition we record the conditions or events that can cause a transition. Below the line we record any action performed during the transition. The at @ symbol in the condition is an event designator and can be read "when . . . ," "on . . . ," or "at the time the event . . . ." The state net in Fig. 1.10 for a lamp thus declares that at the time the event *activate switch* occurs and the lamp is in the *off* state, *apply power to filament* and make the transition to the *on* state. Also, at the time the event *activate switch* occurs and the lamp is in the *on* state, *remove power from filament* and make the transition to the *off* state. State nets may be used to model a wide range of behaviors from this simple example to complex models exhibiting various degrees of concurrent behavior.

Chapter 3 discusses the details of object-behavior modeling using state nets.

Figure 1.9 Lamp in an off state and in an on state.

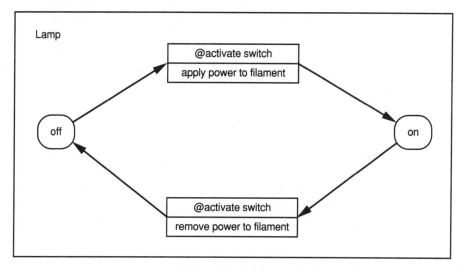

Figure 1.10   A state net.

1.7.3 Large OSA Models

Today's software systems are becoming increasingly large and complex. We cannot expect to use OSA to understand and document large systems unless we provide some additional abstraction and organization features. To solve this problem, we provide views and high-level components.

Views allow analysts to abstract complex object-relationship models and object-behavior models into higher level object-relationship and object-behavior models that present only the most important and most-salient features of the model. These high-level views allow analysts and others to grasp the big picture without getting lost in the details. Since the details may also be important, analysts may expose more and more information about some aspect of a complex model through successively lower level views until every detail has been exposed and explored. This method of abstraction is analogous to the method of packaging object-oriented development modules to hide the internal details of a lower level module from the outside world.

For example, we may consider the object-relationship model in Fig. 1.8 as a high-level view of the objects represented. Vehicles, represented at a high level of abstraction in Fig. 1.8, are complex objects. There are many different kinds of vehicles, and each vehicle is made up of many hundreds of parts and subparts, each of which has a specific behavior. We may hide this large amount of detail under the high-level object class *Vehicle* in Fig. 1.8 and expose it as object-relationship models, at various levels of abstraction, as may be needed.

High-level OSA modeling components include high-level object classes, high-level relationship sets, high-level states, and high-level transitions. Every high-level component is a first-class component and can be used in place of a basic component. When building object-relationship models or object-behavior models, we may start at the most convenient level of abstraction. Often we may wish to start at a high level and then decompose objects into finer and finer detail.

For example, when we first consider lamps, we might wish to just consider the

basic high-level *on* and *off* states as in Fig. 1.10. Later, however, we may be interested
in more detail. If the lamp is a three-way lamp, we may choose to alter the original state
net to show this directly, or we may choose to provide a lower level state net for the *on*
state with the more specific states *dim, medium,* and *bright.*

Chapter 4 discusses the details of the creation and decomposition of views, and
Chap. 5 discusses the details of high-level components.

1.7.4 Object Interaction Models

OSA allows the analyst to express expected interaction among objects in a system. Figure 1.11 shows some of the ways objects commonly interact. In Fig. 1.11(a) a radio
news announcer broadcasts the news on the air waves. Those who wish may tune in
their radio and listen to the news. Figure 1.11(b) shows two people talking on the telephone. Figure 1.11(c) shows a manager giving a work request to a secretary using an
in-basket, which is an intermediate repository. Requests for the secretary go into the in-
basket whenever the manager generates them. Requests are removed from the in-basket
whenever the secretary is ready to work on them.

In OSA, an analyst models interactions using interaction diagrams. Figure 1.12
shows a partial example of how an analyst might model the interactions in Fig. 1.11.
The zigzag arrows are interaction links. They denote the potential for object interaction
and give the direction of the interaction. A description of the interaction is in a label on
the interaction link. We thus see in Fig. 1.12(a) that a *Radio News Announcer* broadcasts news to anyone who wants to listen. In Fig. 1.12(b) we see that a parent and college student are having a conversation, and we see in Figure 1.12c that a manager passes
work requests to a secretary through an in-basket.

The views in Fig. 1.12 are high level. If we want to know more about when the
interactions take place, we can look into the behavior of the objects who are interacting.
For example, if we wish to know when a secretary is ready to receive a work request, we
look at the state net of a secretary. In the state net we would see a transition with an
action such as *get work request,* which defines the point in a secretary's behavior when
the specified interaction takes place.

Chapter 6 discusses the details of object-interaction models.

1.7.5 Model Integration

Model integration provides a way for a team of analysts to create a large OSA model.
When the task is too much for one individual, a group of analysts can divide the task
into individual size pieces. Each individual creates an OSA model or a piece of an OSA
model for part of the larger system. When the OSA models for each part are complete, they
can be integrated to form a single OSA model. The integration process is sometimes
straightforward, but often has some subtleties that make it difficult to merge models.

Chapter 7 points out the difficulties of managing large and complex models and
explains how to create large OSA models through model integration.

1.7.6 The Model Building Process

Building OSA systems-analysis models is not a step-by-step process. Instead, the building activities are usually concurrent, with many interactions among modeling activities.

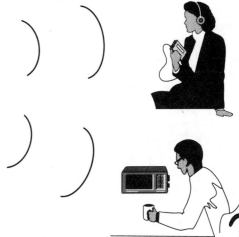

(a)

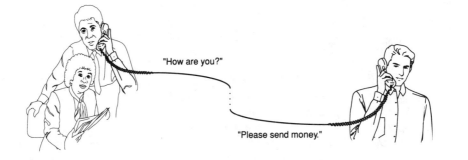

"How are you?"

"Please send money."

(b)

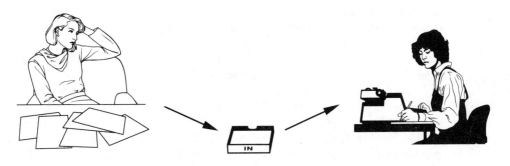

(c)

Figure 1.11 Real-world interaction among objects.

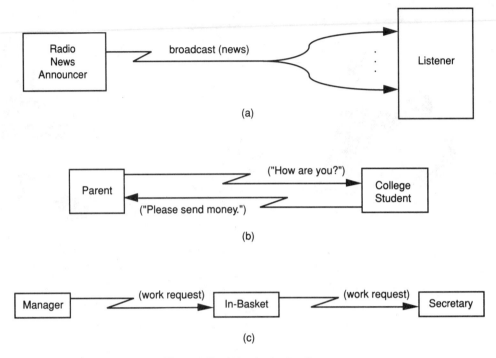

Figure 1.12   Object-interaction diagrams.

We make no strict rules about which modeling activity should take place before another. Once an object class has been identified, an analyst can produce a state net for it before, at the same time, or after the rest of the object-relationship model is developed. Analysts may develop models top-down by decomposition or bottom-up by composition.

Although we do not specify the order in which OSA model components are built, the goal is clear. We wish to create an analysis model that accurately reflects the system being studied and provides the insight and understanding necessary for design and implementation.

In the chapters that follow, we draw examples from many different applications to illustrate the model building process. We also present two reasonably large examples. In Chaps. 2 through 7 we develop an analysis model for a small, fictitious garden-seed company, called the Green-Grow Seed Company. We use this example because it is based on common knowledge about how people work and interact in a business environment. This running example has a data-processing orientation.

We develop the model for our running example bottom-up. In Chap. 2, we develop an object-relationship model for part of the company. In Chap. 3, we develop a state net for an object class. In Chaps. 4 and 5, we create some views for the object-relationship models in Chap. 2 and the state net in Chap. 3, and we do some further development on a state net at a lower level of abstraction. In Chap. 6, we develop much of the object interactions for the Green-Grow Seed Company. Finally, in Chap. 7, we develop and integrate the rest of the object-relationship model and add some additional state nets and interaction diagrams.

By way of contrast, we develop a portion of a satellite tracking system in Chap. 8. This application is real time and scientific. We develop the model top down and interleave our discussion of its object-relationship model, state nets, and communication protocols. The various approaches we take in building these models shows the flexibility and freedom afforded analysts using OSA.

1.8 OSA

OSA (Object-oriented Systems Analysis) is a new approach for capturing and organizing information pertinent to the design and implementation of a software system. The approach places emphasis on system understanding and documentation. Understanding is achieved by building a conceptual model of the system under study. The resulting OSA document provides a consistent foundation for system design and implementation.

1.9 Bibliographic Notes

There are many techniques for performing analysis. A recent book, *Software Requirements: Analysis and Specification*, describes many of the common models and methods [Davis 1990]. It includes a bibliography of 598 references.

Several analysis techniques first appeared in the 1970s. One of the first was the Program Statement Language/Program Statement Analyzer (PSL/PSA) [Teichroew 1977]. At about the same time, Ross developed a Structured Analysis and Design Technique (SADT) [Ross 1977]. In the late 1970s a tool, language, and method, called Software Requirements Engineering Methodology (SREM) was developed to support large systems projects [Alford 1977].

Structured analysis, one of the better known analysis techniques and the foundation for many of the CASE tools, also appeared in the late 1970s. Demarco was the first to produce a book describing structured analysis [DeMarco 1979]. Gane and Sarson later changed some of the representation and added some new concepts [Gane 1979]. In the mid 1980s McMenamin and Palmer enhanced structured analysis by introducing the notion of essential modeling [McMenamin 1984]. Soon after, Ward and Mellor [Ward 1985] and Hatley and Pirbhai [Hatley 1987] extended structured analysis by adding real-time features. In 1989, Yourdon produced a book that summarized the structured analysis approach [Yourdon 1989].

Besides specific analysis techniques, the 1970s also brought us comprehensive development methods that incorporated analysis activities. Because these analysis and design techniques tended to be driven by the structure of data, they became known as data-driven techniques. These techniques include Warnier's Logical Construction of Programs (LCP) [Warnier 1974 and 1981], Orr's Structured Requirements Definition (SRD) [Orr 1981], and Jackson's Jackson System Development (JSD) [Jackson 1983].

In the 1980s, we also saw the emergence of some analysis and specification techniques of a different category. Early in the decade Zave developed an operational approach to system specification called PAISLey [Zave 1982]. At about the same time, another operational approach based on the Gist specification language was also developed [Balzer 1982]. In the mid-80s Wasserman developed the User Software Engineering (USE) method [Wasserman 1986].

Recently, object-oriented analysis models and techniques have appeared. The first to appear was *Object-Oriented Systems Analysis* (OOSA) by Shlaer and Mellor [Shlaer 1988]. This book focuses on the use of ER models for capturing declarative or descriptive information. In 1990, Coad and Yourdon's *Object-Oriented Analysis* (OOA) appeared [Coad 1990]. The Coad-Yourdon approach, like OSA, presents techniques for representing declarative, behavior, and interactive information. Unlike OSA, the information structure and behavior definition of OOA mirrors the classical object-oriented programming paradigm. In 1991, Rumbaugh et al., presented an object-oriented modeling technique (OMT) for analysis [Rumbaugh 1991]. OMT, like OSA, is model-driven. It presents a declarative model represented by an extended ER diagram, a behavior model represented by statecharts [Harel 1988], and a process-interaction model represented by data flow diagrams. The technique also suggests a minimal object-interaction model. The declarative and behavior models are similar to those found in OSA.

2

The Object-Relationship Model

In this chapter we describe the Object-Relationship Model (ORM). As constituent parts of the model we define and discuss objects and relationships, object classes and relationship sets, constraints, and notes. The definitions and discussion are presented informally and include several examples of each major concept.

When using the Object-Relationship Model to do systems analysis for a particular system, we create an ORM model for the system and represent it using an ORM diagram. We have carefully chosen the ORM diagraming conventions. We wish to emphasize, however, that the diagraming conventions are not the model. Rather, the diagrams are the way we represent the model. The object-relationship modeling concepts are independent of the diagraming conventions. Even though other diagraming choices are available, we recommend those that are presented here.

In this chapter, we introduce our running example, the Green-Grow Seed Company. At the end of this chapter, we present a complete ORM for the basic seed company information. We draw from this running example throughout the chapter to illustrate concepts as they are presented.

Our presentation of concepts is bottom up. We first present objects and relationships and then discuss how groups of objects and relationships constitute object classes and relationship sets. Although the presentation is bottom up, we do not mean to imply that analysts should develop ORMs bottom up. Often our ORMs are developed top down by directly conceiving of object classes and relationship sets before becoming concerned about individual objects and relationships.

2.1 Objects

An *object* is a person, place, or thing. An object may be physical or conceptual. Figure 2.1 shows pictures of some sample physical objects: the person *Janet*, the city *New York* the vehicle *Truck #144*, and the *Personal Computer with Serial Number 17844513A*. Each of these objects is a single entity—not a general category of entities. We have *Janet,* not persons, *New York* not cities, and the particular *Personal Computer* whose serial number is *17844513A,* not personal computers in general. The idea is that an object is a single entity or notion. Each object is a unique individual. An object may be related to or made up of other objects, but each object is unique.

An object need not be physical. Figure 2.2 shows some sample conceptual objects: a height of *6'4"*, a temperature of *72°*, and the dollar amount *$100,000*. We may also consider events to be conceptual objects. For instance, we may think of the events *Janet's Graduation* and *The Takeoff of Flight 431* as objects.

In natural language, nouns and noun phrases represent objects. In an ORM diagram, we choose to represent a single object as a dot. To help identify an object, we place one or more nouns or noun phrases by the dot.

Figure 2.3 shows an ORM diagram with some of the objects we have been discussing. Figure 2.4 shows some of the objects we will be discussing for the the the Green-Grow Seed Company. They are *Janet*, *Carol*, and *John* who are employees, *The Manager* who is *Janet, The Green-Grow Seed Company* which is the company being modeled, *12/10/91* which is a shipping date, *#517736* which is an order number, *Best Buy Tomato* which is an item description, *10* which is a quantity ordered, a different *10* which is a quantity shipped, *50 wpm* which is a typing speed, *$20/hour* which is an hourly rate, and *10%* which is a discount rate. In Fig. 2.4 notice that the object *Janet* is the same as the object *The Manager,* and is thus an example of an object with more than one associated noun or noun phrase. Observe also that there are two different objects whose noun is *10*. One is a quantity ordered, and the other is a quantity shipped. "Which is which?" From the ORM diagram in Fig. 2.4, we are unable to tell. We need additional related information to distinguish them as we shall discuss now.

2.2 Relationships

Objects in ORM diagrams are often meaningless unless we understand some relationships among them. The *10*s in Fig. 2.4 are indistinguishable by their name (*10*) alone. As Fig. 2.5 shows, however, if we make one *10* be the number of *Best Buy Tomato* seeds ordered and the other be the number of *Best Buy Tomato* seeds shipped, they become distinguishable and meaningful.

A *relationship* establishes a logical connection among objects. Relationships associate one object with another, similar to the way verbs and verb phrases relate one noun or noun phrase to another. For instance, in the sentence "*Janet is manager of Carol*" the phrase "*is manager of*" specifies a relationship between the objects *Janet* and *Carol*.

In ORM diagrams, we choose to represent relationships with lines, and we provide a name for the relationship. The name is usually a sentence that describes the relationship. We insist that the relationship name includes the object name of each object it connects, and that it includes no other object name. We may thus have the ORM diagram

New York

Janet

PC #17844513A

Truck #144

Figure 2.1   Some sample physical objects.

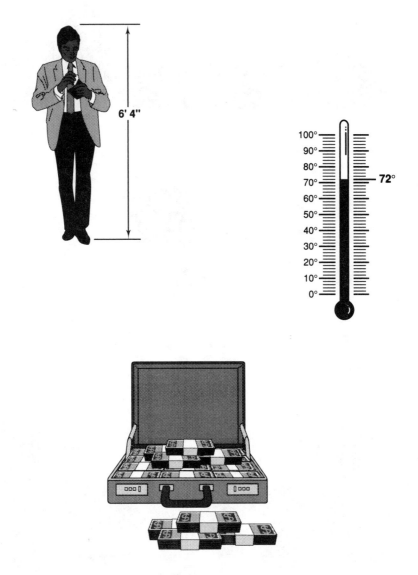

$100,000

Figure 2.2 Some sample conceptual objects.

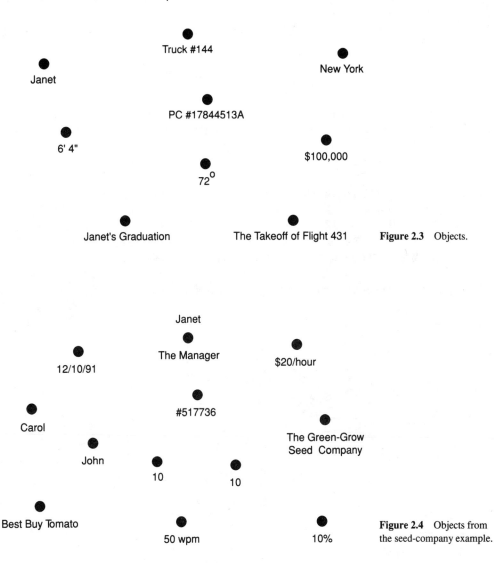

Figure 2.3   Objects.

Figure 2.4   Objects from the seed-company example.

	Packing List	
Description	Quantity Ordered	Quantity Shipped
Best Buy Tomato	10	10
Stand Tall Sweet Corn	30	28

Figure 2.5   Objects in context.

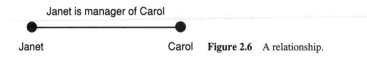

Figure 2.6   A relationship.

shown in Fig. 2.6 for the management relationship of *Janet* and *Carol*. Although relationships are unique, names of relationships are not necessarily unique. We may, for example, have a different *Janet* managing a different *Carol*. Relationships may also have several names. In Fig. 2.6, for example, we may have in addition chosen "*Carol reports to Janet.*"

Other examples of relationships are shown in Fig. 2.7. "*John is 6'4" tall.*" "*Truck #144 has subparts: Body 12245, Engine 2B377681, and Wheel ax11245, Wheel ax11246, Wheel ax11247, and Wheel ax11248.*" "*Personal Computer with Serial Number 17844513A has 5.25 inch Floppy F1227756, 40 Meg Disk H22743A68, and VGA Monitor VGA564431.*" "*Bob repaired Truck #144's Exhaust Pipe.*" "*Susan is married to Bob.*" "*Janet is boss of Janet.*" This last example expresses the notion that *Janet* is her own boss and shows that we may express a relationship of an object with itself.

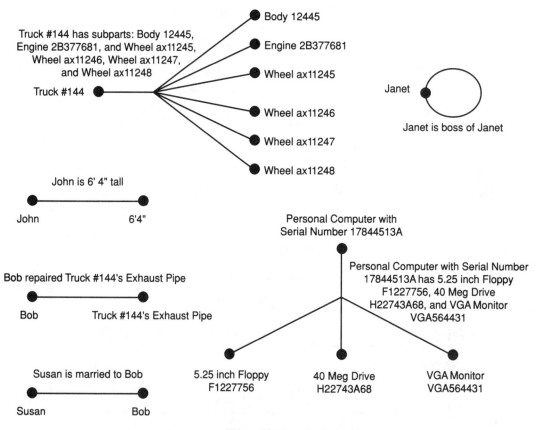

Figure 2.7   Sample relationships.

The number of connections to objects in a relationship is called the *arity* of the relationship. If there are two connections, the relationship is binary. If there are three, the relationship is ternary. For four objects the relationship is quaternary. Relationships with five or more objects are 5-ary, 6-ary, and so on. It is common to refer to relationships of arity three or more as *n-ary* relationships. In Fig. 2.7, we have four binary relationships, one quaternary relationship, and one 7-ary relationship. Even though the *is boss of* relationship in Fig. 2.7 has only one object, it is binary because it has two connections, both to *Janet*. This relationship holds because Janet is her own boss.

Our requirement that a relationship name contain the names of all objects that participate in the relationship helps make relationships understandable, but it also makes names long. We thus provide abbreviations for common cases. For binary relationships we may omit the object names and provide a verb or verb phrase and a directional arrow or bidirectional arrow. We read unidirectional arrows only in the direction of the arrow, and read bidirectional arrows in both directions. Examples of binary relationship abbreviations for the binary relationships in Fig. 2.7 are shown in Fig. 2.8. Since these are abbreviations, we fill in object names when we express the relationships. We thus have "*John has height 6'4"*," "*Susan is married to Bob,*" "*Bob is married to Susan,*" "*Bob repaired Truck #144's Exhaust Pipe,*" and "*Janet is boss of Janet.*" Observe that there are two names for the marriage relationship of *Bob* and *Susan*. We read bidirectional relationships once in one direction and once in the other direction.

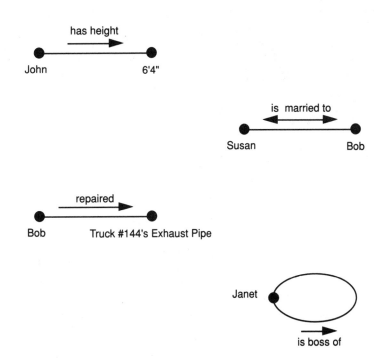

Figure 2.8   Relationship name abbreviations.

2.3 Object Classes

Identifying and documenting individual objects and relationships among objects is useful, but very tedious and not powerful enough for documenting most systems. Analysts need organizational techniques to manage the large number of objects and relationships in today's complex systems. To manage this complexity we need some method of abstracting and grouping a large body of facts into smaller, more comprehensible units.

Identification of sets of objects that belong together for some logical reason is called *classification*. In OSA, a set of objects that belong together for some logical reason is called an *object class*. The Object-Relationship Model encourages analysts to organize objects into object classes. Each object class has a name that is generic and denotes any member of the object class. Thus, in an ORM, an object class whose name is *X* designates a classification of objects each of which is considered to be an *X*. Since each object in object class *X* is an *X*, the objects in the class are alike, at least in this sense.

For example, the object class *Person* for the Green-Grow Seed Company is the set of persons of interest to the company. The objects *Janet, Bob, Carol*, and *John* may all be members of the *Person* object class, since each one is a person.

Objects grouped together in an object class usually have other common traits besides just being class members. The idea of an object class is similar in concept to the technique of biological classification, where biologists group living things together that share common traits. An analyst may group any set of objects into an object class for any reason, but the classification should make good sense. Otherwise, the analyst will have difficulty communicating about the classification, just as a biologist is likely to have trouble communicating about an arbitrary collection of living things.

In an ORM diagram we choose to represent an object class by a rectangle enclosing its name. Figure 2.9 shows several sample object classes, some of which will become part of the ORM for the Green-Grow Seed Company. If *Janet* and *Carol* are employees of the company, they are members of the *Employee* object class. If *Janet, John*, and *Carol* are customers of the company, they are members of the *Customer*

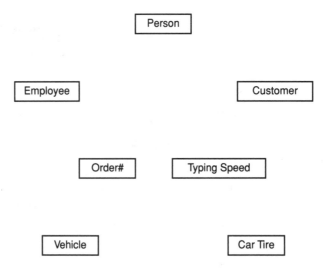

Figure 2.9   Sample object classes.

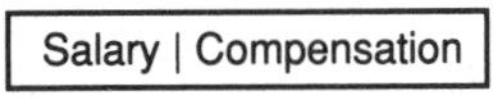

Salary | Compensation

Figure 2.10   An object class with multiple names.

object class. Vehicles such as *Truck #144* and Jon's *Sports Car (VIN 444XYZ)* could be members of the *Vehicle* object class.

The name for an object class should be chosen such that it can be applied to a single object. *Janet* is a *Person,* not a *Persons. Truck #144* is a *Vehicle,* not a *Vehicles.* Object class names should be specific and well qualified. *Car Tire* would be better than just *Tire* if there is any possibility of confusion in the analysis, say, with bicycle tires.

At times, we need more than one name for an object class. For instance, an analyst may have interviewed two different organizations of the same company in which they use different names for the same concept. In an ORM we choose to write all names in the rectangle representing the object class and separate multiple names by a vertical bar (|). Figure 2.10 shows the use of the multiple naming feature for a *Salary | Compensation* object class. We may call each object in the class a salary or a compensation.

As an object changes over time, the object classes to which it belongs may change. For example, a person enters the *Employee* object class when becoming employed, enters the *Salaried Employee* object class upon being promoted to a manager, and leaves the *Salaried Employee* object class at retirement.

2.4 Relationship Sets

Along with encouraging the organization of objects into object classes, the Object-Relationship Model also encourages analysts to organize relationships into relationship sets. Although we allow arbitrary classification of objects, so long as they are meaningful to the analyst, we are more restrictive for relationship sets. We require that individual relationships of a relationship set correspond in a particular way.

We illustrate the particular correspondence we require in Fig. 2.11, which shows five pairs of relationships. Intuitively, the two relationship names in each pair express the same logical connection among the objects. In each pair, observe that the number of objects referenced by the relationship is the same, three in Fig. 2.11(e) and two in all the others, and that the object names are in the same place in the relationship name. In Fig. 2.11(a), for example, the object names *Jon* and *Phil* are at the beginning of the relationship name and *Sports Car (VIN 444XYZ)* and *Truck #144* are at the end. The corresponding non-object-name phrases are not only in the same place, but are also identical. In Fig. 2.11(e), for instance, the non-object-name phrases are "earns" and "as of," which both occur in the same place in the relationship names.

The corresponding objects in each pair of relationships in Fig. 2.11 appear to belong to the same object class. Indeed, we require that they do. For Fig. 2.11, we assume that objects *Phil, Jon, Janet, Bob, Susan,* and *John* are members of the *Person* object class, *Truck #144* and *Sports Car (VIN 444XYZ)* are members of the *Vehicle* object class, *The Green-Grow Seed Company* and *ZZY Inc.* are members of the *Company* object class, *$50,000* and *$79,000* are members of the *Salary* object class, and *1 Sep 89* and *1 Sep 91* are members of the *Salary Effective Date* object class.

The correspondence we require for relationships in a relationship set may be thought of as a template with object classes designating slots for objects and phrases that

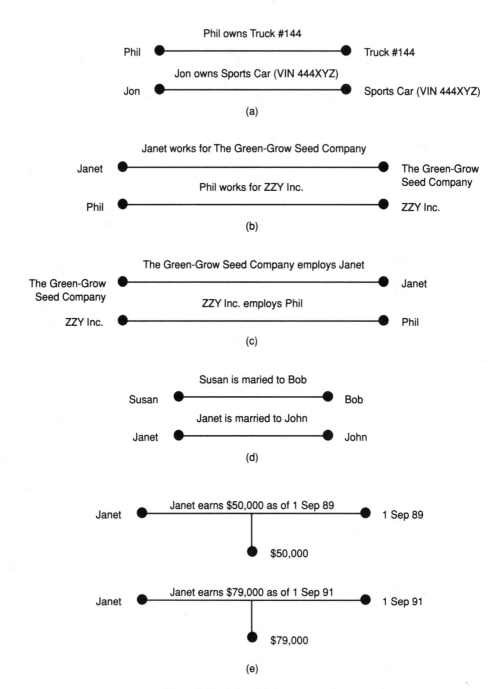

Figure 2.11 Pairs of similar relationships.

express a logical connection among the objects. Figure 2.12 gives a template for each pair of relationships in Fig. 2.11. A *relationship set* is a set of relationships where each matches the same template and has as its name the text of the matched template. Where different names imply the same logical connection between objects, there may be more than one matched template and thus more than one name for a relationship set.

Our definition of a relationship set guarantees that all relationships in a relationship set have the same arity, all connect to the same object classes, and all express the same logical connection among the objects. Since, in practice, we usually specify a relationship set first and thus first specify a template before worrying about any member relationship, we are able to easily satisfy the relationship-set rule for names.

Figure 2.13 shows an ORM diagram representation of relationship sets for the pairs of relationships in Fig. 2.11. There are only four diagrams for the five pairs because the pairs in Figs. 2.11(b) and 2.11(c) express the same relationships and are thus both represented by Fig. 2.13(b).

As a basic graphical representation for all relationship sets, we have chosen a diamond with lines connecting associated object classes. There is a line for each connection to an object class. Because we must have a connecting line for each object occurrence in a relationship set, there are two lines to *Person* in Fig. 2.13(c), not one. We complete the notation by writing one or more relationship set names near the diamond. Each name is the text of a template matching the relationship set.

Relationship set names may become lengthy. For binary relationship sets, which in practice constitute the majority of all relationship sets, we have chosen a shorthand

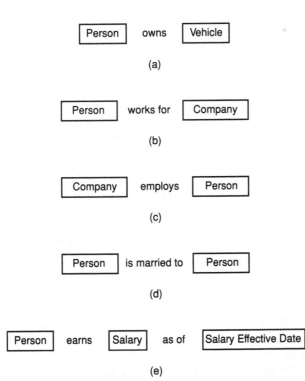

Person   owns   Vehicle

(a)

Person   works for   Company

(b)

Company   employs   Person

(c)

Person   is married to   Person

(d)

Person   earns   Salary   as of   Salary Effective Date

(e)

Figure 2.12   Relationship-set templates.

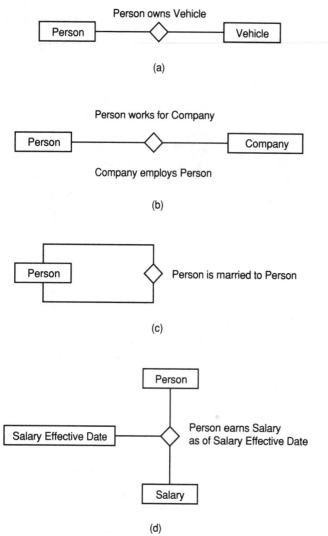

Figure 2.13   Relationship sets for the pairs of relationships in Fig. 2.11.

notation. Most names for binary relationship sets naturally consist of one phrase with one object class name before the phrase and one after the phrase. If the name of a binary relationship does not have this form, we can always rewrite it so that it does. In the shorthand notation for binary-relationship names, we omit the diamond, omit the object class names, and add a directional arrow. In the shorthand notation, we thus represent the relationship set in Fig. 2.13(a) as shown in Fig. 2.14(a) and in Fig. 2.13(b) as shown in Fig. 2.14(b).

Notice that we read the relationship set name *employs* in Fig. 2.14(b) from right to left as the arrow dictates rather than from left to right. When creating ORM diagrams it is often convenient to be able to have binary relationships read from any direction, even on a diagonal. The arrow directs the proper interpretation of the relationship set name.

The shorthand notation for Fig. 2.13(c) is more interesting. Figure 2.14(c) shows a bidirectional arrow. We read bidirectional arrows in both directions. Thus, we would

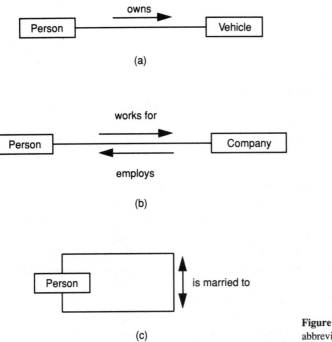

(a)

(b)

(c)

Figure 2.14   Relationship sets with abbreviated names.

not only have the relationship "*Susan is married to Bob*" as given in Fig. 2.11(d) but also, for the same relationship, "*Bob is married to Susan.*" For this particular example the relationship-set name "*Person is married to Person*" reads identically in both directions, but this is an anomaly of the example, not a feature of bidirectional arrows. We see this in Fig. 2.15 where the bidirectional arrow reads "*Spouse is married to Employee*" in one direction and "*Employee is married to Spouse*" in the other direction.

We provide no shorthand notation for *n*-ary relationship sets. Thus, we leave the relationship set in Fig. 2.13(d) as it is.

An object class is often connected to more than one relationship set. In Fig. 2.16, for example, *Person* connects to both "*Person owns Vehicle*" and "*Person works for Company.*" When we build ORM diagrams, we are usually interested in many different object classes and their connecting relationship sets. Objects in an object class may participate in all relationship sets to which the class is connected.

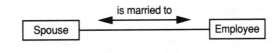

Figure 2.15   Bidirectional relationship set.

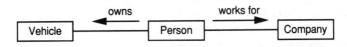

Figure 2.16   An object class with two relationship sets.

2.5 Constraints

We can model much of a system with objects, relationships, object classes, and relationship sets. To describe a system more satisfactorily, however, we often wish to state additional properties of object classes and relationship sets by imposing constraints.

For instance, we might want our model to allow only one height for a person or at most one object in a *President* object class when modeling a particular company.

In these examples the constraints restrict the conditions under which an object can be considered to be a member of an object class or the conditions under which a relationship can be considered to be a member of a relationship set. In an ORM, a *constraint* restricts the membership of one or more object classes or relationship sets.

The Object-Relationship Model allows analysts to express several different types of constraints. Cardinality constraints include object-class cardinality constraints to restrict the number of objects in an object class and participation constraints and co-occurrence constraints to restrict relationship sets with respect to objects in connecting object classes. Special relationship sets including generalization, specialization, aggregation, and association impose restrictions directly on relationship sets and indirectly on object classes. Special object classes including singleton object classes and relational object classes impose restrictions directly on object classes and indirectly on relationship sets. In addition, the Object-Relationship Model also has general constraints that allow analysts to express arbitrary conditions that restrict object classes and relationship sets.

We continue in this section by discussing participation constraints, co-occurrence constraints, and object-class cardinality constraints. In subsequent sections, we discuss special relationship sets, special object classes, and general constraints.

2.5.1 Participation Constraints

A *participation constraint* defines the number of times an object in an object class can participate in a connected relationship set. We require that every connection of a relationship set to an object class have a participation constraint.

The basic form for a participation constraint is a pair: *min:max*. The *min* is a nonnegative integer, and the *max* is either a nonnegative integer or a star (*). The star designates an arbitrary nonnegative number greater than the minimum. In an ORM diagram we place a participation constraint near the connection of the object class and relationship set to which it applies.

For instance, Fig. 2.17(a) shows the participation constraint *0:1* near *Employee,* which thus constrains an employee to participate in the "*Employee has Vacation Schedule*" relationship set at most one time. Similarly, the *1:1* participation constraint near *Vacation Schedule,* constrains a vacation schedule to participate once and only once. Figure 2.17(b) shows the constraint that a person has one and only one height as a *1:1* participation constraint near the *Person* object class. The *0:** participation constraint near *Height* allows a single height object to be associated with zero or more persons. Heights not associated with any person may, for example, be associated with an inanimate object such as a piece of equipment.

Figure 2.18 shows two binary relationship sets with objects, represented as dots, and individual relationships, represented as lines between the dots. These examples

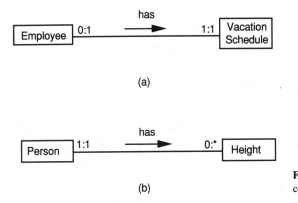

(a)

(b)

Figure 2.17 Sample participation constraints.

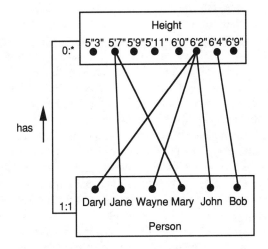

(a)

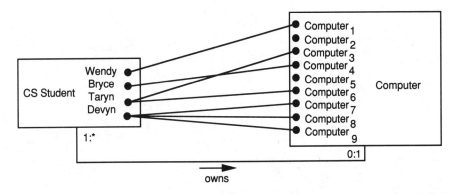

(b)

Figure 2.18 Interpretation of participation constraints.

show the meaning of the common participation constraints *0:1, 1:1, 0:***, and *1:***. Figure 2.18(a) shows that each person in the *Person* object class participates once and only once in the "*Person has Height*" relationship set. Thus, for every dot representing a person in the *Person* object class there is one and only one relationship that connects it to a dot in the *Height* object class. Figure 2.18(a) also shows that a height can be connected to zero or more persons. Thus, every dot in the *Height* object class has zero or more lines emanating from it. The participation constraint *1:*** in Fig. 2.18(b) declares that every CS student must own at least one computer, but can own many. Thus, every dot in the *CS Student* object class has at least one line emanating from it; some have several. The *0:1* participation constraint in Fig. 2.18(b) declares that a computer has at most one CS-student owner. Thus, every dot in the *Computer* object class has at most one line emanating from it; some have none.

While the participation constraints *0:1, 1:1, 0:***, and *1:*** are the most common, others are often useful. For instance, Fig. 2.19 shows a *2:2* participation constraint on the *Person* object class, which constrains each person to participate exactly twice in the "*Person has Parent*" relationship set and thus to have exactly two parents.

Instead of using just nonnegative integers, we may also use variables in participation constraints. Figure 2.20 illustrates the use of participation constraints with variables. The *a:a* participation constraints on *Athlete*, in both the "*Athlete plays for Team*" relationship set and the "*Athlete participates in Team Sport*" relationship set, constrain an athlete to participate exactly *a* times in both relationship sets. Thus, the constraints ensure that the number of teams for which an athlete plays is the same as the number of team sports in which the athlete participates. An athlete may not, for example, play on two teams and participate in three team sports or participate in two team sports and play on three teams. Since *a* is variable, however, it can be different for different athletes, and thus one athlete may play on two teams and participate in two team sports while another athlete plays on three teams and participates in three team sports.

The relationship set in each of the examples we have shown so far is binary. Participation constraints also apply to *n*-ary relationship sets. Figure 2.21 shows the "*Person earned Degree on Date*" relationship set with participation constraints. In *n*-ary relationship sets participation constraints have exactly the same meaning as in binary relationship sets. In Fig. 2.21, a person participates between zero and three times and

| Person | 2:2 | has | 1:* | Parent |

Figure 2.19   A useful 2:2 participation constraint.

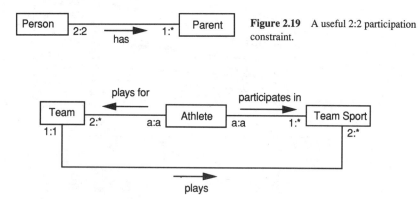

Figure 2.20   Variables in participation constraints.

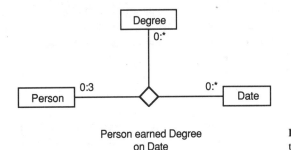

Person earned Degree
on Date

Figure 2.21 Participation constraints in a ternary relationship set.

thus may have up to three different degrees recorded in this system. A degree participates zero or more times. Thus, there may be degrees that no one has earned, and there may be degrees that many people have earned. A date also participates zero or more times. There may be days on which no one earned a degree, and there may be days on which many people earned degrees.

We provide a shorthand notation for participation constraints. The notation is applicable when the minimum and maximum values of a participation constraint are identical. For example, if a participation constraint is *1:1*, the shorthand notation for this constraint is simply *1*. If the constraint is *a:a*, the shorthand notation is *a*. Figure 2.22 is a version of Fig. 2.20 using the shorthand notation for applicable participation constraints.

Participation constraints may be generalized to allow multiple basic constraints for each connection between an object class and relationship set. Figure 2.23 gives an example of generalized participation constraints. The ORM in Fig. 2.23 declares that each *Model 32A Computer* must be configured with *0, 1, 2,* or *4* four-megabyte memory expansion modules. Any other configuration would not meet the requirements declared by this analysis. We see from the example diagram that participation constraints are

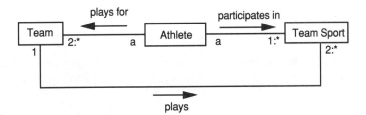

Figure 2.22 Shorthand participation constraints.

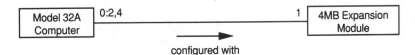

Figure 2.23 Generalized participation constraints.

generalized by listing each basic constraint and separating each constraint by a comma. Such a designation declares a set of possible participation values. Each object in the object class must participate *n* times where *n* is one of the possible participation values in the designated set.

2.5.2 Co-occurrence Constraints

Participation constraints restrict the number of relationships in which an object can appear in a relationship set. We may also wish to restrict the number of different objects that can appear together with a particular object or a particular group of objects in a relationship set. Co-occurrence constraints allow us to make these restrictions.

Consider, for example, the *"Student received Grade in Course taken during Semester"* relationship set in Fig. 2.24. All of the participation constraints are *0:**. A student participates once for each course taken, and thus zero times for students who have not yet completed any course and many times for students who have completed many courses. A course participates once for each student in every semester the course is offered, and thus zero times for courses that have not gone through one offering and many times for courses that have been offered several times. A grade is given to each student for each course, and thus participates zero times in the rare case that no student has received a particular grade and many times for most grades. A semester participates once for each course taken by a student during the semester, and thus zero times for future semesters and many times for past semesters. Since all the participation constraints are *0:**, the participation constraints do not restrict the relationship set.

There are, however, some restrictions we might like to express. For example, a student cannot have more than one grade for an offering of a course during a semester. If students cannot repeat courses, the restriction is even tighter, since then a student cannot have more than one grade for a course. There may also be some lower limits on the number of students in a course and on the number of courses a student can take in any one semester.

We can use co-occurrence constraints to express these restrictions. The first co-occurrence constraint in Fig. 2.24 is

$$\textit{Student} \;\; \textit{Course} \;\; \textit{Semester} \;\xrightarrow{\;1:1\;}\; \textit{Grade}$$

which ensures that there is one and only one grade for a student in a course during a particular semester. The second co-occurrence constraint in Fig. 2.24 is

$$\textit{Course} \;\; \textit{Semester} \;\xrightarrow{\;5:*\;}\; \textit{Student}$$

which declares the minimum course size for a semester to be five. The last co-occurrence constraint in Fig. 2.24 is

$$\textit{Student} \;\; \textit{Semester} \;\xrightarrow{\;1:12\;}\; \textit{Course}$$

which limits the number of courses a student can have taken for a particular semester to twelve.

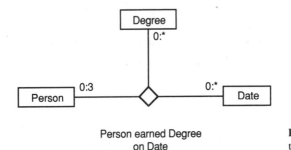

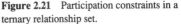

Person earned Degree
on Date

Figure 2.21 Participation constraints in a ternary relationship set.

thus may have up to three different degrees recorded in this system. A degree participates zero or more times. Thus, there may be degrees that no one has earned, and there may be degrees that many people have earned. A date also participates zero or more times. There may be days on which no one earned a degree, and there may be days on which many people earned degrees.

We provide a shorthand notation for participation constraints. The notation is applicable when the minimum and maximum values of a participation constraint are identical. For example, if a participation constraint is *1:1*, the shorthand notation for this constraint is simply *1*. If the constraint is *a:a*, the shorthand notation is *a*. Figure 2.22 is a version of Fig. 2.20 using the shorthand notation for applicable participation constraints.

Participation constraints may be generalized to allow multiple basic constraints for each connection between an object class and relationship set. Figure 2.23 gives an example of generalized participation constraints. The ORM in Fig. 2.23 declares that each *Model 32A Computer* must be configured with *0, 1, 2,* or *4* four-megabyte memory expansion modules. Any other configuration would not meet the requirements declared by this analysis. We see from the example diagram that participation constraints are

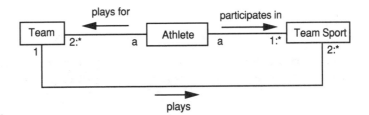

Figure 2.22 Shorthand participation constraints.

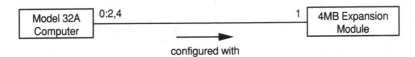

Figure 2.23 Generalized participation constraints.

generalized by listing each basic constraint and separating each constraint by a comma. Such a designation declares a set of possible participation values. Each object in the object class must participate n times where n is one of the possible participation values in the designated set.

2.5.2 Co-occurrence Constraints

Participation constraints restrict the number of relationships in which an object can appear in a relationship set. We may also wish to restrict the number of different objects that can appear together with a particular object or a particular group of objects in a relationship set. Co-occurrence constraints allow us to make these restrictions.

Consider, for example, the *"Student received Grade in Course taken during Semester"* relationship set in Fig. 2.24. All of the participation constraints are $0:*$. A student participates once for each course taken, and thus zero times for students who have not yet completed any course and many times for students who have completed many courses. A course participates once for each student in every semester the course is offered, and thus zero times for courses that have not gone through one offering and many times for courses that have been offered several times. A grade is given to each student for each course, and thus participates zero times in the rare case that no student has received a particular grade and many times for most grades. A semester participates once for each course taken by a student during the semester, and thus zero times for future semesters and many times for past semesters. Since all the participation constraints are $0:*$, the participation constraints do not restrict the relationship set.

There are, however, some restrictions we might like to express. For example, a student cannot have more than one grade for an offering of a course during a semester. If students cannot repeat courses, the restriction is even tighter, since then a student cannot have more than one grade for a course. There may also be some lower limits on the number of students in a course and on the number of courses a student can take in any one semester.

We can use co-occurrence constraints to express these restrictions. The first co-occurrence constraint in Fig. 2.24 is

$$\textit{Student Course Semester} \xrightarrow{\textit{1:1}} \textit{Grade}$$

which ensures that there is one and only one grade for a student in a course during a particular semester. The second co-occurrence constraint in Fig. 2.24 is

$$\textit{Course Semester} \xrightarrow{\textit{5:*}} \textit{Student}$$

which declares the minimum course size for a semester to be five. The last co-occurrence constraint in Fig. 2.24 is

$$\textit{Student Semester} \xrightarrow{\textit{1:12}} \textit{Course}$$

which limits the number of courses a student can have taken for a particular semester to twelve.

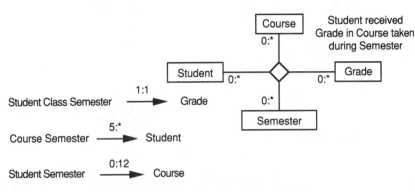

Figure 2.24   Sample co-occurrence constraints.

A *co-occurrence constraint* specifies the minimum and maximum number of times an object or combination of objects can co-occur in the relationships of a relationship set with another object or combination of objects. In the co-occurrence constraint

$$1\text{:}12$$
$$Student \;\; Semester \longrightarrow Course$$

in Fig. 2.24, a particular combination of student-semester objects can co-occur with between one and twelve different course objects. We can therefore, for example, have the combination of objects *Janet-Fall92* appearing in the relationship set with at most twelve different courses.

Figure 2.25(a) shows a violation of the ORM co-occurrence constraint

$$1\text{:}1$$
$$Student \;\; Course \;\; Semester \longrightarrow Grade$$

Janet cannot have two different grades for *CS100* for the same semester. Figure 2.25(b) shows a valid relationship set for this co-occurrence constraint. *Janet* may repeat *CS100* in a different semester and get a different grade. Although not illustrated in Fig. 2.25(b), *Janet* can even get the same grade for the same course in two different semesters. The co-occurrence constraint only declares that for each student-course-semester triple there is one and only one grade, and not that for each student-course-semester triple there is a different grade.

Notationally, we have chosen to write co-occurrence constraints by listing one or more object-class names on the left and right of a right-pointing arrow and by providing a list of *min:max* pairs above the arrow. A *max* may be a star, and if the minimum and maximum are the same value, we may use a shorthand notation similar to participation constraints where only a single number is written above the arrow. We may also use variables as we did in participation constraints. On an ORM diagram, we place a co-occurrence constraint close to the relationship set to which it applies.

To properly form co-occurrence constraints, several restrictions must be observed. (1) The object-class names in the left and right lists of a co-occurrence constraint must be a subset of the object-class names in the relationship set to which the co-occurrence constraint applies. (2) Neither the left list nor the right list may be empty. (3) An object-class name may not be listed more than once in a co-occurrence constraint. (4) Minimums cannot be zero because any combination of left objects either does not appear in a relationship set or appears at least once with some combination of right

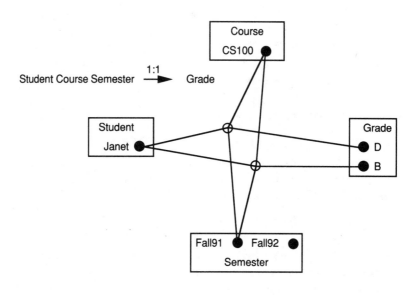

(a)

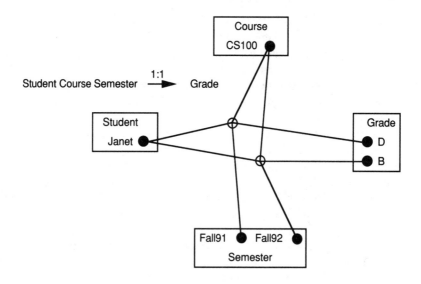

(b)

Figure 2.25 Invalid and valid relationship sets for a co-occurrence constraint.

objects. For example, an attempt to use *0:12* for the co-occurrence constraint for limiting the number of courses per semester to twelve makes no sense because a student who has zero courses during a semester cannot participate in a relationship in the "*Student received Grade in Course taken during Semester*" relationship set for that semester.

Figure 2.26 shows an example of a co-occurrence constraint for The Green-Grow Seed Company. The company has a long-held policy of giving a bonus to employees in their paycheck near the anniversary of their date of hire as a reward for good job performance. The co-occurrence constraint

$$Employee \longrightarrow Work\ Anniversary$$

declares that an employee has only one work anniversary. A co-occurrence constraint with no given numeric constraint denotes *1* or *1:1* by default. The example also illustrates that for co-occurrence constraints we count the occurrences of objects in relationship sets with respect to their association with other objects, not just occurrences. An employee may have received many bonuses from the company, perhaps one each year for the past several years. In each of these relationships, the employee and work anniversary are the same. The relationship set does not violate the co-occurrence constraint, which merely declares that there is only one work anniversary per employee, and not that we can only record the employee-work anniversary pair once in the relationship set.

2.5.3 Object-Class Cardinality Constraints

So far we have considered that each object class may contain an unlimited number of members. In many cases, however, the number of objects in an object class may be restricted. This restriction applied to an object class is called an *object-class cardinality constraint*.

We write object-class cardinality constraints in the same format as participation constraints and place them inside the top right-hand corner of the object-class rectangle. Shorthand notation may also be used so that if the minimum and maximum values are identical, only a single number or variable need be shown. When an object-class cardinality constraint is not specified, we assume the constraint to be *0:**.

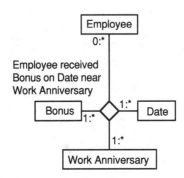

Figure 2.26   A co-occurrence constraint in the seed-company example.

Figure 2.27 An object-class cardinality constraint.

Figure 2.27 shows an example of an object-class cardinality constraint. There are exactly 50 US states. Hence, the object-class cardinality constraint for the *State in the USA* object class is *50*.

2.6 Special Relationship Sets

Several types of relationship sets appear so often in ORM diagrams that we have chosen special symbols to represent them. These include the *is a* relationship set, the *part of* relationship set, and the *is member of* relationship set. We discuss each of these in the sections that follow.

2.6.1 Generalization—The *Is A* Relationship Set

If we have object classes *Person* and *Student,* we can use an *is a* relationship set to state that every student is a person. Since every student is a person, the set of objects in object class *Student* is a subset of the set of objects in object class *Person*. If there are persons of interest to the system who are not students, then object class *Student* is a proper subset of object class *Person*. Whether a proper subset or not, the essential idea for *is a* relationship sets is that one object class is a subset/superset of another.

A superset object class is called a *generalization,* and a subset object class is called a *specialization. Person* is a generalization of *Student,* and *Student* is a specialization of *Person*.

In an ORM diagram we choose to represent an *is a* relationship set with a transparent triangle. We connect a generalization object class to a vertex of the triangle and a specialization object class to the base opposite the connecting vertex. Figure 2.28 shows an example. The generalization *Person* connects to a vertex, and the specialization *Student* connects to the opposite base. The orientation of the triangle does not

Figure 2.28 A sample generalization-specialization.

matter. A generalization object class may be above, below, or beside a specialization object class.

We do not label an *is a* relationship set because the transparent triangle tells us to read the relationship set as *is a,* in the direction from specialization to generalization. We thus read the name of the relationship set in Fig. 2.28 as *"Student is a Person."*

An *is a* relationship set has participation constraints, but we usually omit them. Participation constraints on a specialization are always *1:1* since every specialization object is a generalization object. The default participation constraints on the generalization are *0:1*. The only other choice is *1:1,* which happens only when the set of objects in a generalization is always identical to the set of objects in a specialization. A *1:1* participation constraint on *Person* in Fig. 2.28 would be appropriate, for example, if the only persons of interest in the system were students.

Figure 2.29 shows another example. Here we have more than one specialization for a generalization. This relationship set is read, *"Student is a University Person, and Instructor is a University Person, and Staff Member is a University Person."* An object in a generalization can be a member of zero or more specializations. A university person, for example, might be both a student and an instructor. A staff member might take a course and might even also teach some other course and therefore be in all three specialization as well as the generalization. A full-time research professor is a university person, but might not be classified in any of the specializations shown in the diagram. The diagram in Fig. 2.29 could have been written as three separate *is a* relationship sets, but it is often more convenient to write specializations of a common generalization using only one triangle.

2.6.1.1 Specialization Constraints.   Besides being more convenient, we are sometimes able to be more expressive when we group specializations under a generalization. Sometimes we wish to state that together the objects in the specializations constitute all the objects in the generalization. We may also wish to state that the specializations are mutually exclusive, or that they form a partition of the generalization. We provide notation for these types of specialization constraints as explained in the following paragraphs.

A *union constraint* declares that the union of a group of specializations of a generalization constitutes the entire membership of the generalization. When we have a union constraint, we know that every member of a generalization is a member of at least

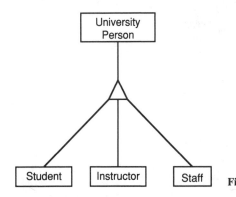

Figure 2.29 Multiple specializations.

one specialization in the group. We show an example of a union constraint, along with its representation, in Fig. 2.30. The union symbol (∪) inside the triangle represents the union constraint and ensures that a university person is either a student, a faculty member, or a staff member. There are no other kinds of university persons in the system. A university administrator, for example, would have to be considered either as a faculty member, staff member, or student. The union constraint does not preclude a member of one specialization from being a member of another specialization. For example, a person who is a member of the faculty may also be a student at the university.

A *mutual-exclusion constraint* declares that a group of specializations of a generalization are pairwise disjoint. We show an example of mutual exclusion, along with its representation, in Fig. 2.31. The partitioning-plus symbol (+) in a triangle designates that the specialization object classes in the group are pairwise disjoint. In the diagram, this means that no 80XXX-based PC is a 680XX-based PC and vice versa. Since no union constraint was applied there may be, however, PCs that are neither 80XXX-based nor 680XX-based. These would be in the *PC* object class, but neither in the *80XXX-based PC* nor the *680XX-based PC* object class.

A *partition constraint* declares that a group of specializations partitions a generalization. A partition requires that the partitioning sets be pairwise disjoint and that their union constitute the partitioned set. We thus have both a union constraint and a mutual-exclusion constraint and thus choose to denote a partition constraint by a union symbol combined with a partitioning-plus symbol. Figure 2.32 shows a partitioning of the *Employee* object class of the Green-Grow Seed Company. The diagram declares that each member of the *Employee* object class must belong to one and only one of the object classes *Manager, Packager,* or *Clerk.* For example, an *Employee* cannot be both a *Manager* and a *Clerk.*

2.6.1.2 Roles.   To better understand relationship sets, we sometimes label the connections of object classes in relationship sets with role names. For instance, in the relationship set connecting *Employee* to *Vacation Schedule* in Fig. 2.33(a) we label the connection to *Employee* as *Full-Time Employee.*

Roles are a short way to express specializations. The ORM diagram in Fig. 2.33(a), for example, is a more concise representation of the ORM diagram in Fig. 2.33(b). In Fig. 2.33(b) we have created an object class for the role *Full-Time Employee*, and we have made the new class a specialization of *Employee*. In addition,

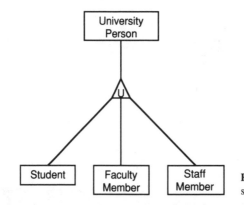

Figure 2.30 A union constraint on specializations.

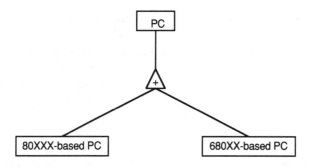

Figure 2.31 A mutual-exclusion constraint on specializations.

we have shifted the relationship so that it is between *Full-Time Employee* and *Vacation Schedule* rather than between *Employee* and *Vacation Schedule*. We have also changed the participation constraints so that every *Full-Time Employee* has a *Vacation Schedule*.

Since roles are shorthand notation for diagraming specialization classes, we read the relationship names as if the diagram were in the expanded form. Thus, we read the relationship in Fig. 2.33(a) as "*Full-Time Employee has Vacation Schedule*."

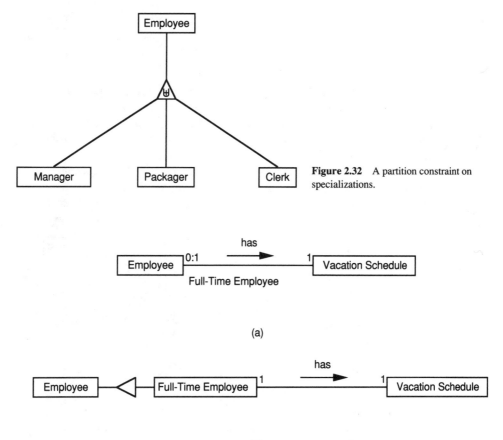

Figure 2.32 A partition constraint on specializations.

(a)

(b)

Figure 2.33 Roles as specializations.

We require that the minimum participation constraint for a role object class be at least one. This forces all members of the specialization role object class to participate in the relationship set at least once and allows us to properly interpret participation constraints where roles are involved. In Fig. 2.34, for example, an employee of the Green-Grow Seed Company need not have any dependents, but a breadwinner always has at least one dependent. The subset of employees who have dependents are precisely those who are in the *Breadwinner* specialization object class designated by the role.

A *role,* marking a connection, is thus a specialization of the object class in the connection whose objects fully participate in the relationship set in the connection. The only difference between the generalization and the role specialization is the mandatory participation imposed on members of the specialization. No other implicit or explicit constraints distinguish specialization classes from roles.

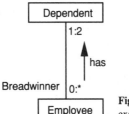

Figure 2.34 A role in the seed-company example.

2.6.1.3 Inheritance.

Since the members of a specialization are also members of a generalization, they all participate in the relationship sets of the generalization. The converse, however, does not hold because some members of a generalization may not be members of a specialization and may, therefore, not participate in relationship sets of the specialization. This observation leads to the notion of inheritance.

As Fig. 2.35 shows, both the *Employee* object class and the *Customer* object class in the Green-Grow Seed Company are specializations of *Person*. We notice from the participation constraints in the diagram that every *Person* has a name and an address.

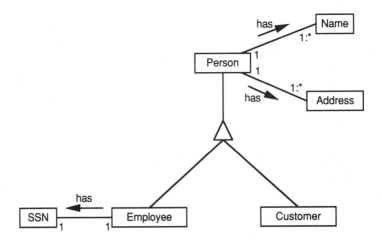

Figure 2.35 Inheritance of relationship sets from a generalization.

Since every employee is a person and every customer is a person every employee and every customer also has a name and address. A specialization *inherits* the relationship sets of its generalization. Because of inheritance, we do not have to re-establish those relationship sets for a specialization. They are implied by the *is a* relationship set.

Since all members of a generalization are not necessarily members of a specialization, a generalization does not inherit the relationship sets of a specialization. In Fig. 2.35, the *"Employee has SSN"* relationship set is not inherited by the object class *Person*. While it is true that some persons have social security numbers, it is not necessary that all persons have social security numbers. All persons who are employees, however, have social security numbers.

2.6.1.4 Multiple Inheritance.   In an ORM, we have *multiple inheritance* when a class is a specialization of two or more generalizations. Figure 2.36 gives an example of an object class *Student Newspaper Editor,* which is a specialization of both the *Student* object class and the *Newspaper Editor* object class. The diagram shows that there is a class of newspaper editors and a class of students, and that all student newspaper editors are both students and newspaper editors. Members of the object class *Student Newspaper Editor* inherit from both the *Newspaper Editor* object class and the *Student* object class. They, therefore, attend a school and started editing a newspaper on some date. Student newspaper editors must also be registered for a journalism class, but neither students in general nor newspaper editors in general need be registered for a journalism class. The meaning of the diagram would also be the same if we had used two specialization relationships with two triangles rather than one grouping triangle.

Another way of viewing multiple inheritance is to see the set of objects in the specialization object class as a subset of the intersection of two or more generalization classes. The specialization having multiple inheritance need not be a proper subset of the intersection, but each member of the specialization must be a member of the intersection.

When the subset is exactly the intersection, we may use an intersection symbol (∩) in a grouping triangle to constrain the subset to be the intersection. Figure 2.37 shows an example that is appropriate for the Green-Grow Seed Company. A customer-employee is both a customer and an employee, and every employee who is a customer is a member of the *Customer-Employee* object class.

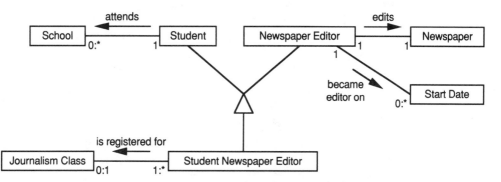

Figure 2.36 Multiple inheritance.

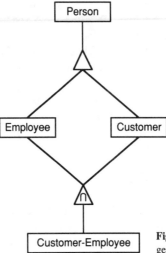

Figure 2.37 Intersection of multiple generalizations.

2.6.2 Aggregation—The *Is Part of* Relationship Set

Another type of relationship set that appears often is the *is part of* relationship set. This relationship set declares that an object, called a *superpart* or *aggregate,* is composed of other objects called *subparts* or *components.*

Figure 2.38 shows an example of an aggregation relationship set. The representation we have chosen for an *is part of* relationship set is a solid-filled triangle. This differs from the representation for an *is a* relationship set which is a transparent triangle instead of a solid-filled triangle. The aggregate object class connects to the vertex of the triangle, and the component object classes connect to the opposite base.

We read *is part of* relationship sets the same as we read *is a* diagrams. For instance, we read the diagram in Fig. 2.38 as a *"Body is part of Car, and Engine is part of Car, and Wheel is part of Car."*

Aggregation relationship sets have participation constraints. If an aggregate object class has more than one component object class (the usual case), then it is difficult to write a participation constraint next to the aggregate object class and be able to properly associate it with a component object class. To alleviate this problem, we write participation constraints below the black triangle as shown in Fig. 2.38. The participation constraints of Fig. 2.38 declare that every car must have one and only one body and engine and exactly four wheels and that every body, engine, and wheel is part of one car or not part of any car.

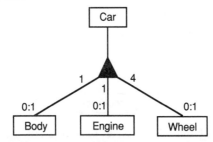

Figure 2.38 A sample aggregation.

An aggregation does not require that all subparts be represented. Thus, a reader of an ORM should not assume that the collection of subparts (from the object classes connected to the base of the triangle) constitute the whole (in the object class connected to the vertex of the triangle). In Fig. 2.38, a body, an engine, and four wheels may or may not be all that is required to constitute a car.

Since there is often more than one way to divide an aggregate into components, there may be several *is part of* relationship sets for one aggregate object class. Figure 2.39, for example, shows that we can view a software division of a company as two different aggregations. Here, we decompose a software division into an analysis section, a design section, a coding section, and a testing section, and we also decompose it into a business section, a science section, and a systems software section.

Figure 2.40 shows an example of an aggregation for the Green-Grow Seed Com-

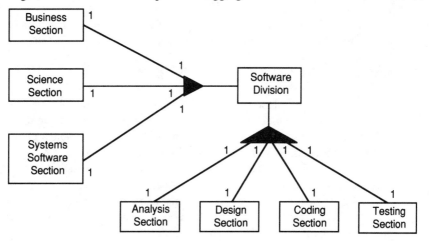

Figure 2.39   An aggregation with two different decompositions.

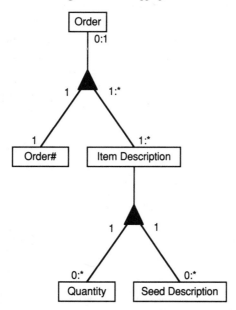

Figure 2.40   An aggregation for the Green-Grow Seed Company.

pany. The diagram shows that orders are composed of a single order number and one or more item descriptions. Each item description, in turn, is composed of a quantity and a seed description. The participation constraints further stipulate that an order number uniquely identifies an order, that one or more item descriptions appear on an order, and that quantities and seed descriptions may appear in zero or more different item descriptions.

2.6.3 Association—The *Is Member of* Relationship Set

The last type of relationship set that has a special representation is the *is member of* relationship set. We use the *is member of* relationship set to form a set of objects that we wish to consider as a single object. Here, the group formed is a set instead of an assembly.

Figure 2.41 shows an example. A group of students may join together in an association to form a student club, such as a chess club, a drama club, or a science club. In Fig. 2.41 members of the *Student* object class are students whereas each member of the *Student Club* object class is a set of students. The connecting relationship set maintains information about which students are members of which clubs.

Is member of relationship sets are always binary. The objects of one of the connecting object classes are set objects whose members are objects in the other connected object class. In Fig. 2.41 the elements in the *Student Club* object class are set objects whose members come from the *Student* object class.

The object class whose objects are sets is called the *set class* or the *association*, and the object class whose objects are members is called the *member class* or the *universe*. *Association* denotes that the members associate together to form an object. *Universe* is an often-used mathematical notion that denotes the set of objects from which subsets may be formed. The set class in Fig. 2.41 is *Student Club,* and the member class is *Student*. Each club is an association and is a subset of the universe of students.

The set class may contain more than one set with the same membership. In our school club example, a physics club and a chemistry club might consist of the same group of students. Each object in a set class has its own identity independent of the members that belong to it.

Notationally, we have chosen to represent an *is member of* relationship set by a line with a star on the connection to the set class. The relationship set is read from the member side to the set side. Thus, in Fig. 2.41 we read the relationship set as *"Student is member of Student Club."*

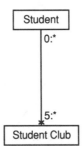

Figure 2.41 Sample association.

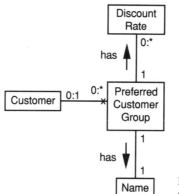

Figure 2.42 An association for the Green-Grow Seed Company.

Participation constraints have their usual meaning. In Fig. 2.41, the *0:** participation constraint means that a student need not be a member of any club, but may be a member of many clubs. The *5:** participation constraint means that every student club at the school must have at least five members.

Figure 2.42 shows an example of the use of an association for the Green-Grow Seed Company. The diagram declares that a preferred customer group is composed of customers. The participation constraints say that a customer need not belong to any preferred customer group and may belong to at most one. They also say that a preferred customer group may contain zero or more customers. The zero minimum value allows a preferred customer group to exist even if it has no members.

2.7 Special Object Classes

The Object-Relationship Model has two special types of object classes. One is a singleton object class which has exactly one member. The other is a relational object class whose members are relationships.

2.7.1 Singleton Object Classes

Sometimes we may be interested in modeling object classes that always have exactly one object. For the Green-Grow Seed Company we have two examples. One is the company itself. There is only one company of interest for this system. The other example is the manager of the Green-Grow Seed Company. The company is small and has only one manager. The manager may change from time to time, but there is one, and only one.

We can, of course, model these singleton objects as objects without object classes. We can even have relationships among objects, but what do we do if we wish to relate an object with an object class? In the ORM for the Green-Grow Seed Company, for example, we wish to state that the company employs three or more employees and that every employee of interest works for the company.

According to our definitions, a relationship only connects individual objects and a relationship set only connects object classes. From this definition we would have trouble showing the relationships between employees abstracted by the employee object class and the Green-Grow Seed Company. To solve this problem we have borrowed an idea from physics. Depending on their objective, physicists view light as a wave or as a composition of particles. In an similar manner, we may view a dot on an ORM diagram either as a representation for a single object or as a representation for a special object class called a singleton object class, depending on our objective as systems analysts.

A *singleton object class* is an object class that contains one and only one object. Since there is only one object in the object class, all the properties of the object class are properties of the single object. The object-class name is the object name. If there are several object-class names, they are all names for the single object.

Figure 2.43 shows an example of a singleton object class. Here, we treat the object *The Green-Grow Seed Company* as an object class. We are thus able to establish the relationship set "*Employee works for The Green-Grow Seed Company.*" Using participation constraints, we are also able to say that there are at least three employees and that every employee of interest works for the company.

Figure 2.43   A sample singleton object class.

Figure 2.44 shows an alternate notation for the same example. Here, we show the object as an object class, but add the object-class cardinality constraint *1* to designate that the object class is a singleton object class. There is no semantic difference between the diagrams in Fig. 2.43 and 2.44.

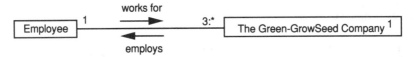

Figure 2.44   A singleton object class shown as a rectangle.

Although there is no semantic difference, there is a reasonable connotative difference. When a singleton object class is represented as a rectangle, it looks like a container restricted to have one object. This suggests that the contents of the rectangle may change during the lifetime of the system. On the other hand, when a singleton object class is represented as a dot, it looks fixed, which suggests that it is not likely to change over the life of the system.

The Green-Grow Seed Company, for example, has one manager, but the manager changes from time to time. For instance, *Janet* may be the manager today, but *Phil* might succeed her and become the manager later. For this reason, we prefer to represent the object class for the manager as a rectangle as Fig. 2.45 shows. The name of the

Figure 2.45   Singleton object class whose content changes over time.

object class is *Manager*, not *Janet,* because we are thinking of the manager as a position in the company, not the person who is currently the manager.

In addition to providing a way to connect objects to object classes, the singleton object class also provides a way to extract and address an individual member of an object class. Figure 2.46 shows the *Person* object class and the object *Janet*. If Janet is a person, we may show the singleton object class *Janet* as a specialization of *Person*. We can thus show both an object class and enumerate the members of the object class in an ORM diagram.

2.7.2 Relational Object Classes

So far, we have treated relationship sets and object classes as if they were entirely different concepts. Sometimes, however, we wish to treat relationships in a relationship set as objects in an object class. For instance, we may wish to think of the relationships in the relationship set *"Female Person is married to Male Person"* as marriages. Each relationship is a marriage object. We may further wish to relate marriage objects to other objects such as marriage dates and wedding licenses.

In an Object-Relationship Model we have chosen to represent relationships in a relationship set as objects of an object class by drawing a rectangle that includes the relationship set and its associated object classes and placing a new object-class name in the rectangle. An object class whose members are relationships is called a *relational object class*.

Figure 2.47 shows an example of a relational object class. The relationships in the relationship set *"Female Person is married to Male Person"* constitute the relational object class called *Marriage*. The *Marriage* object class is an object class just like any other object class. It may, for example, participate in relationship sets such as the *"Marriage has Marriage Date"* and *"Wedding License documents Marriage"* relationship sets shown in Fig. 2.47. In this case, the objects of the marriage class that participate in these relationship sets are themselves relationships, but are abstracted as objects.

Figure 2.46 Class membership for an individual object.

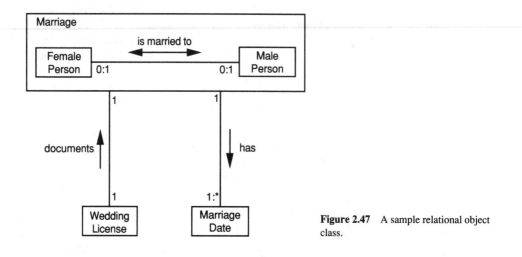

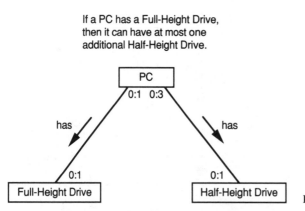

Figure 2.47   A sample relational object class.

2.8 General Constraints

Participation constraints, co-occurrence constraints, object-class cardinality constraints, special relationship sets, and special object classes cannot represent all the constraints an analyst might want to impose in an ORM. We thus provide general constraints as part of the Object-Relationship Model. A *general constraint* is a statement that restricts the membership of one or more object classes or relationship sets.

Figure 2.48 shows an example. The participation constraints declare that a personal computer (PC) can have up to three half-height drives and one full-height drive. The following general constraint, which is also part of the diagram in Fig. 2.48, further restricts the two *has* relationship sets.

If a PC has a Full-Height Drive, then it can have at most one additional Half-Height Drive.

Thus, although the participation constraints allow zero or one full-height drive and zero, one, two, or three half-height drives, the general constraint prohibits both the combina-

Figure 2.48   A sample general constraint.

tion of a full-height drive and two half-height drives and the combination of a full-height drive and three half-height drives. On the other hand, either two or three half-height drives with no full-height drive is permissible because it satisfies both the participation constraints and the general constraint.

Users of the Object-Relationship Model may write general constraints as they wish. A formalist may prefer some variation of predicate-calculus notation, whereas others may prefer a concise natural language characterization.

Figures 2.49, 2.50, and 2.51 show three additional examples of general constraints for The Green-Grow Seed Company. In Fig. 2.49, the constraint

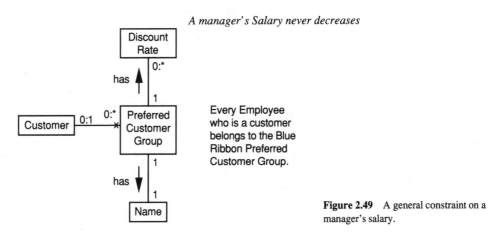

A manager's Salary never decreases

Every Employee who is a customer belongs to the Blue Ribbon Preferred Customer Group.

Figure 2.49 A general constraint on a manager's salary.

restricts the relationship set *"Manager earns Salary as of Salary Effective Date."* For successive salary-effective dates, the salary value must be greater than or equal to the previous salary value. In Fig. 2.50, the constraint

Every Employee who is a customer belongs to
the Blue Ribbon Preferred Customer Group

restricts both the *Preferred Customer Group* object class and the *"Customer is member of Preferred Customer Group"* relationship set. The constraint strengthens the *0:1* par-

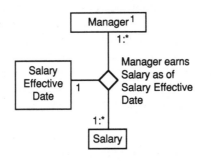

A Manager's Salary never decreases.

Figure 2.50 A general constraint on customer groups.

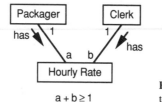

Figure 2.51 General constraints applied to participation constraints.

ticipation constraint on the relationship set to *1:1*, but only for employees who are customers. The constraint does not restrict the *Employee* object class, but does depend on it.

In Fig. 2.51, the general constraint

$$a + b \geq 1$$

refers to participation constraints. It restricts the *Hourly Rate* object class and the two *has* relationship sets with respect to each other. Each hourly rate must participate at least once in the two *has* relationship sets. This constraint declares that we are interested in only those hourly rates assigned to the company's packagers or clerks.

We may check the satisfaction of the general constraint by substituting, for each object in *Hourly Rate*, the number of times it participates in "*Packager has Hourly Rate*" for *a* and the number of times it participates in "*Clerk has Hourly Rate*" for *b*, adding, and checking that the sum is greater than or equal to one.

2.9 Membership Conditions

To define the common criteria that determines set membership for object classes and relationship sets, we introduce membership conditions. A *membership condition* for an object class is a conjunction of all the constraints imposed by an ORM on the object class. Similarly, a *membership condition* for a relationship set is a conjunction of all the constraints imposed by an ORM on the relationship set. An object that satisfies the conditions for membership in an object class is a member of the object class, and a relationship that satisfies the conditions for membership in a relationship set is a member of the relationship set.

To illustrate membership conditions, we first give in Fig. 2.52 the complete, basic ORM for the Green-Grow Seed Company. Most of the object classes and relationship sets that appear have been previously discussed. In what follows, we will discuss much of the diagram again, from the point of view of membership conditions.

Object-class membership conditions describe objects. For an object, they give the object-class name as the class of the object, the generalizations of which the object is a subclass, the direct relationships, the inherited relationships, and the constraints applicable to the object including participation constraints, co-occurrence constraints, constraints that apply because of special object classes and relationship sets, and any applicable general constraints.

Consider, for example, the membership condition for the object class *Order*. The membership condition for an object *X* to be in the class *Order* is a conjunction of the following constraints taken from Fig. 2.52.

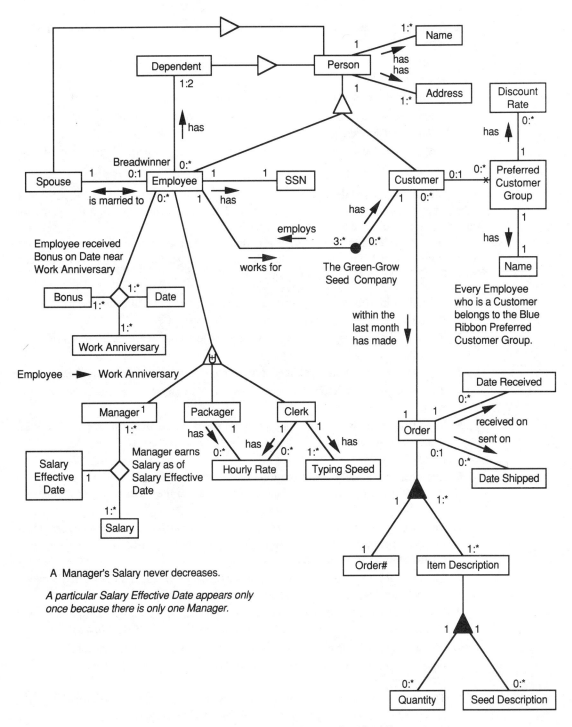

Figure 2.52   Basic ORM for the Green-Grow Seed Company.

1. *X* is an order.

2. *X* appears once as an order in the relationship set "*Order received on Date Received.*"

3. *X* appears once as an order in the relationship set "*Customer within the last month has made Order.*"

4. *X* appears once as an order in the relationship set "*Order# is part of Order.*"

5. *X* appears one or more times as an order in the relationship set "*Item Description is part of Order.*"

6. *X* appears at most one time as an order in the relationship set "*Order sent on Date Shipped.*"

Membership conditions for an object class help the reader of an ORM determine what a particular object in the object class is. The membership condition just given for the *Order* object class tells us what an order is.

As another example, we derive a textual characterization of the manager of the Green-Grow Seed Company from the ORM. Again, we take the conditions directly from Fig. 2.52, but this time we list them more informally.

1. An object in *Manager* is a manager (because of the classification constraint that declares that for object class *C* every object in *C* is a *C*).

2. There is one manager (because of the object-class cardinality constraint on *Manager*).

3. The manager is an employee (because *Manager* is a specialization of *Employee*).

4. The manager is a person (because *Manager* is indirectly a specialization of *Person*).

5. The manager earns a salary as of a salary-effective date (because *Manager* participates at least once in the relationship set "*Manager earns Salary as of Salary Effective Date*").

6. The manager's salary information for various salary-effective dates is known (because *Manager* may participate several times in the relationship set "*Manager earns Salary as of Salary Effective Date*").

7. The manager works for the Green-Grow Seed Company (because *Manager* inherits participation in the "*Employee works for The Green-Grow Seed Company*" relationship set from *Employee*).

8. The manager has a social security number (because *Manager* inherits participation in the "*Employee has SSN*" relationship set from *Employee*).

9. The manager may be married to a spouse (because *Manager* inherits participation in the "*Employee is married to Spouse*" relationship set from *Employee*).

10. The manager may be a breadwinner (because *Manager* inherits participation from *Employee* in the *Breadwinner* role of the relationship set "*Breadwinner has Dependent*").

11. The manager may have dependents (because *Manager* inherits participation in the "*Breadwinner has Dependent*" relationship set from *Employee*).

12. The manager has a name (because *Manager* inherits participation in the "*Person has Name*" relationship set from *Person*).

13. The manager has an address (because *Manager* inherits participation in the "*Person has Address*" relationship set from *Person*).

14. The manager may be a customer (because a manager is in the *Employee* object class which may have a non-empty intersection with the *Customer* object class).

15. The manager is neither a packager nor a clerk (because the object classes *Manager, Packager,* and *Clerk* partition the object class *Employee*).

16. The manager's salary never decreases (as directly stated in a general constraint).

17. If the manager is a customer, then the manager is a member of the Blue Ribbon preferred customer group (because a manager is an employee and because of the general constraint, "*Every Employee who is a customer belongs to the Blue Ribbon Preferred Customer Group*").

Anyone who manages the Green-Grow Seed Company must satisfy all 17 of these conditions.

Membership conditions for relationship sets are usually simpler. They consist of the template that stipulates which relationships may belong to the set, associated participation and co-occurrence constraints, and any applicable general constraints. For example, a relationship in the "*Employee received Bonus on Date near Work Anniversary*" relationship set in Fig. 2.52 must satisfy the following conditions.

1. The relationship name must have the form "*E received B on D near WA*" where *E* is in object class *Employee, B* is in *Bonus, D* is in *Date,* and *WA* is in *Work Anniversary*.

2. An employee may have received several bonuses on different dates near the work anniversary (because participation constraints allow *E, B, D,* and *WA* to appear several times).

3. An employee has one and only one work anniversary (because co-occurrence constraint *Employee $\longrightarrow$ Work Anniversary* forces every *E-WA* pair that appears in the relationship set to be identical).

2.10 Notes

Analysts may wish to add comments, explanations, pictures, or sketches to an ORM. A *note* adds information to an ORM, but does not restrict any object class or relationship set. A note, for example, may record who created an ORM diagram and the date on which it was last revised, it may explain a part of the diagram that may otherwise appear unusual, or it may provide a photograph or sketch of some object modeled in the ORM.

As an example, Fig. 2.52 contains one note:

A particular Salary Effective Date appears only once because there is only one Manager.

This note reminds those who think the participation constraint on *Salary Effective Date* should have been *1: **, that this could only be true if there were more than one Manager. A manager does not get two raises on the same day.

Since notes do not constrain object classes or relationship sets, an ORM diagram has the same meaning with, or without them. The added information, however, is often useful.

2.11 Bibliographic Notes

The Object-Relationship Model allows an analyst to record information about objects and their relationships. Process-oriented techniques either do not provide a means for recording ORM-like declarative information [Jackson 1983] or provide only data-structure techniques such as data dictionaries [DeMarco 1979, Gane 1979], Warnier/Orr diagrams [Orr 1981], and extended declarative structures [Davis 1977, Ross 1977, Teicheroew 1977].

To capture more declarative information, newer analysis techniques and CASE tools are moving away from data-structured models towards relational [Wasserman 1986] and semantic data models [McMenamin 1984, Ward 1985, Yourdon 1989]. All the object-oriented techniques use semantic data models [Coad 1991, Rumbaugh 1991, Shlaer 1988].

Our Object-Relationship Model is also a semantic data model. Recent surveys that define and illustrate semantic data models can be found in *ACM Computing Surveys* [Hull 1987, Peckham 1988].

Semantic data models have their roots in the Entity-Relationship Model [Chen 1976]. Many of the recent semantic data models have embraced the abstraction mechanisms used in OSA, including generalization, specialization, aggregation, and association [Rumbaugh 1991].

We have omitted attributes as a basic construct of ORMs because we consider the declaration of attributes at analysis time to be premature and potentially harmful to the analysis process [Embley 1991b]. It is widely held that the notion of an attribute is hard to define and that it is particularly difficult to distinguish between objects and attributes [Goldstein 1989]. Forced designation of attributes during analysis leads to unnecessarily complex constructs such as multivalued and composite attributes [Ling 1985] and weak entities [Chen 1976]. Forced designation of attributes during analysis also disallows the possibility of expressing relationships among attributes and causes later problems for model integration [Batini 1986] and data normalization [Teorey 1986]. Furthermore, later definition of attributes during design can be done automatically using a process similar to synthesis for relational database systems [Embley 1989].

Relationship sets in ORMs are similar to relationship sets in most semantic models. Some models, however, allow only binary relationship sets [Mark 1983], and most other models do not allow relationships to be expressed among individual object instances.

Typical cardinality constraints are one-one, one-many, and many-many [Chen 1976]. Instead of adopting these typical constraints, we have chosen to use more powerful cardinality constraints [Embley 1991c] including participation constraints [Abrial 1974, Elmasri 1985], and co-occurrence constraints. Co-occurrence constraints allow us to express functional dependencies directly as in relational database design [Codd 1970, Codd 1972], and also allow us to generalize functional dependencies by expressing minimums and maximums.

In OSA generalization and specialization differ somewhat from the earliest presentations of generalization and specialization [Smith 1977], but correspond to more-recent views [Kung 1990, Kung 1989]. Inheritance in OSA also differs. In many object-oriented semantic models, an object class inherits only attributes [Coad 1991, Rumbaugh 1991, Shlaer 1988]. In an ORM, the objects in a specialized object class inherit participation in all relationship sets connected to any generalization.

2.12 Exercises

Throughout the book, we shall use a fictitious public monetary banking system, called the KWE Bank (The Kurtz-Woodfield-Embley Bank), as a basis for our exercises. Below we provide a partial natural language specification for the KWE bank. Portions of this specification will be used in the following exercises. Other portions will be used in exercises in later chapters.

The KWE Bank.   The KWE Bank is composed of one or more branches and a main office. Branches are geographically separate banking offices. All business transactions conducted at each branch are treated by the main office as transactions of the overall KWE banking system. All customers at a branch are also treated as customers of the overall banking system.

Bank Personnel.   Each branch has one and only one branch manager. A branch manager supervises two to three mortgage officers, one to two personal loan officers, and three to five full-time tellers. The main office has a bank president with two secretaries and three vice presidents, each with one secretary. Employees can have only one job in the banking system.

Customer.   Any person who has one or more open accounts in the KWE banking system is considered to be a customer. Persons who do not have open accounts may also be considered as customers if they make use of other services in the banking system. A customer has a name, a mailing address, a social security number, an optional daytime phone number, and an optional evening phone number. A customer must have either a daytime phone number or an evening phone number. All of the customer information is maintained at the main office of the KWE Bank.

Signature Card.   Each customer who opens an account at a branch fills out a signature card, to be kept on file at that branch. The KWE Bank requires one signature card per customer at each branch where a customer has open accounts. In addition to a signature, each card includes the date signed, the name of the customer, a mailing address, a social security number, an optional daytime phone number, and an optional evening phone number. Signature cards are used to validate customer signatures and customer information for transactions (e.g., deposits, withdrawals) against an account.

Account.   An account is associated with a specific branch. An account is either a checking account or a savings account. No other account category exists. All accounts have a balance, an account number, and a date (which is the date on which they were opened). Each savings account has a single interest rate which is fixed on the first business day of each month. A checking account may have one overdraft protection amount which may be changed once during any business day.

Bank Card.   The KWE banking system issues bank cards to some of its customer for use at automated teller machines (ATMs). These ATMs must be connected to an ATM network that has access to the KWE bank ATM system. Bank cards are issued to customers, allowing them to perform deposit, withdrawal, or balance inquiry transactions on an account through an ATM. A bank card has a name and a card number. The name is the name of the customer to which the card was issued. The card number is associated with one or more account numbers, referring to accounts that the card holder may access. Each card number has a personal identification number (PIN) that, when entered by the user, authorizes access to the customer's accounts.

Automated Teller Machine (ATM).   The components of the ATM include a user interface module (UI module). The UI module has a button group and a display. The buttons in the group include transaction selection buttons, numeric-entry buttons, and a cancel button. The transaction selection buttons include four buttons, identified as button one through button four. The numeric-entry buttons include ten numeric buttons and one enter button. The display has a prompt region and four labels for the buttons named label one, label two, label three, and label four. The ATM also has a cash drawer, a card reader, a transaction printer, and an envelope slot. Each ATM has a unique network identification number for communication with the bank computing system. Also, each ATM is assigned to a single branch for service and maintenance responsibility. However, not all branches have ATMs to which they are assigned.

New Year's Club.   The KWE Bank has a New Year's Club. All customers who join receive a bonus interest rate for maintaining a qualifying balance in one of their savings accounts. The qualifying balance must be maintained from the first business day of the calendar year to the last business day of the calendar year.

2.1 The ORM diagrams in Figs. 2.53 and 2.54 show two ways to model relationships among accounts, customers, and bank cards.
   a. Give relationship instances allowed by the ORM in Fig. 2.53 that would not make sense according to our specification for the KWE bank as given. Explain how to disallow these relationship instances.
   b. Give relationship instances allowed by the ORM in Fig. 2.54 that would not make sense according to our specification for the KWE bank as given. Use co-occurrence constraints to disallow these relationship instances.
   c. Discuss the practicality of ensuring correct relationships among objects using participation constraints alone.

2.2 Develop a model showing the relationships among accounts and objects directly related to accounts. In your model, show that checking accounts and savings accounts are account specializations.

2.3 Using the natural language specification as a basis, model an ATM as an aggregation. Model also all objects directly related to an ATM or its component parts.

2.4 If there are n branches, what are the minimum and maximum number of mortgage officers? Give an object class for mortgage officers with an object-class cardinality constraint that limits the number of mortgage officers to this range.

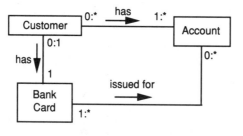

Figure 2.53   Binary relationships among account, customer, and bank card.

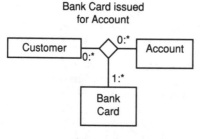

Figure 2.54   Ternary relationship among account, customer, and bank card.

2.5 Develop an ORM diagram showing the employee hierarchy for the branches of the KWE banking system. Also include objects directly related to employees.

2.6 Model the concept of the New Year's Club and document the requirements for a customer to belong to the club using an ORM diagram. As part of this ORM, make New Year's Club an association of New Year's Club members.

2.7 Model the main-office staff for the KWE banking system. Include the KWE Bank as a singleton object class in the model.

2.8 Model the vice presidents and their secretaries as a relational object class of work partners. Show that it is possible to model concepts such as work schedules and date the working relationship began as object classes related to work partners.

3

The Object-Behavior Model

The objective of behavior modeling is to understand and document the way each object in a system interacts, functions, responds, or performs. A behavior model for an object is similar to a job description for an object. Both describe what an object must do within a system.

When we model the behavior of an object in OSA, we record its perceived states, the conditions and events that cause it to change from one state to another, the actions it performs, and the actions performed on it. We may also record exceptions to normal behavior and real-time constraints imposed on object behavior.

We use state nets to document object behavior. In doing so, we use our preferred symbols for visual presentation. We recommend these symbols, but again we hope the reader will look beyond the symbols and concentrate on the underlying behavior modeling foundation of OSA.

We again use the Green-Grow Seed Company as a running example. Although we draw examples from the behavior of several different objects associated with the company, we focus in particular on the behavior of a packager. We give examples from the packager state net throughout the chapter. At the end of the chapter we give a complete state net for packager behavior.

3.1 States

A basic part of behavior modeling is the set of states an object exhibits in a system. In OSA, a *state* represents an object's status, phase, situation, or activity.

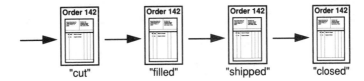

Figure 3.1   States of a purchase order object.

There are several ways to acquire knowledge about an object's state. Usually we observe or imagine an object in a system and mentally record its various states. For example, we recognize that an automobile engine is in a "running" state because we hear its internal components moving. We know that a checking account is "overdrawn" because its current balance is negative. We recognize that an airplane is "taxiing" when we see it moving along the ground.

Figure 3.1 shows a set of states for a purchase order object from the Green-Grow Seed Company. We consider a purchase order to be *cut* when a company clerk assigns a number to a customer's request for seeds. We consider it *filled* after a packager places all items called for in a box and completes and encloses the invoice form. An order is *shipped* after the packaged items have been sent to the customer. Finally, an order is *closed* when a clerk records that the order has been paid for in full.

3.2 Triggers and Transitions

While identification of object states is an essential part of behavior modeling, it is of little value without examining how one state relates to another in the dynamic context of system operation. Specifically, we must understand the expected events and system conditions that activate a change of state for an object.

The process of changing the state of an object is called a *transition*. The events and conditions that activate state transitions are called *triggers*.

Figure 3.2 shows how we can enhance the behavior description of a purchase order by adding a description of the triggers to which the purchase order responds. We

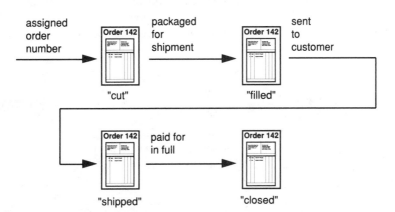

Figure 3.2   Triggers for state transitions of a purchase order.

show trigger statements in Fig. 3.2 as text above a transition arrow. When an object is in a state at the tail of a transition arrow and the trigger associated with the arrow holds, the object leaves the state at the tail and enters the state at the head of the arrow. Thus, for example, when a *filled* order is *sent to customer,* the order transitions into the *shipped* state.

3.3 Actions

In addition to states and transitions among states, we also wish to model the actions an object performs. An *action* may cause events, create or destroy objects and relationships, observe objects and relationships, and send or receive messages.

We place actions into two categories in OSA: *noninterruptible* actions and *interruptible* actions. *Noninterruptible* actions are actions that the analyst expects to run to completion unless exceptions or system failures occur. *Interruptible* actions may be suspended before they finish executing and may resume execution at a later time. In OSA, we think of actions associated with transitions as noninterruptible, whereas actions associated with states are interruptible.

We illustrate these concepts by considering the behavior of a robot used to transport seed containers for the Green-Grow Seed Company. Figure 3.3 shows a robot's transition from the *idle* state to the *moving to container* state. To make the transition, the robot must perform the action comprised of *release wheel brake* and *engage forward motor.* Notationally, we record an action below a line under the transition's trigger. Figure 3.3 shows the action under the trigger *receives transfer-container command.* This action should always complete unless the robot's brake or gear shift fails.

An example of an interruptible action is modeled by the *moving to container* state in Fig. 3.4. When the *arrives at container station* trigger becomes true, the robot should leave the *moving to container* state and enter the *securing container* state. Figure 3.4 shows the action required for this state transition recorded under the line beneath the trigger. A robot must *disengage forward motor, apply wheel brake,* and *activate arm* before entering the *securing container* state.

We commonly use states to model a nondiscrete or long-running activity, or a continuous activity that must be interrupted to allow a response to events or conditions that affect an object. Our robot, for example, moves continuously until it arrives at the container station.

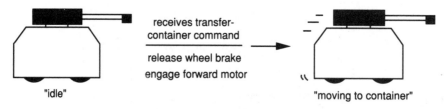

Figure 3.3   Noninterruptable action performed during robot state transition.

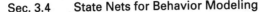

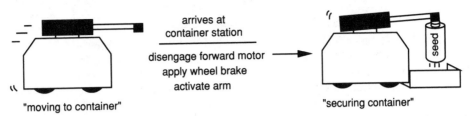

arrives at
container station
───────────────────
disengage forward motor   ⟶
apply wheel brake
activate arm

"moving to container"   "securing container"

Figure 3.4   Interrupting the *moving to container* action of the manufacturing robot.

3.4 State Nets for Behavior Modeling

In OSA we use state nets to model object behavior. A *state net* is a configuration of symbols representing states and state transitions for all objects in a particular object class. We consider a state net to be a "behavior template" that specifies the expected behavior for instances of an object class. When we associate a state net with an object class, we declare that every instance of the object class has the behavior described by the state net.

3.4.1 State Modeling

In state nets we choose to represent states by rounded rectangles in state nets. Each rounded rectangle corresponds to a single state of an object in the context of the system under study. State names are inside the rounded rectangle representing the state. Figure 3.5 shows a partial state net for our manufacturing robot. The state net declares some of the robot's states.

To associate a particular state-net diagram with an object class, we place the state-net diagram in a rectangle representing the object class and write the object-class name near the top left-hand corner of the rectangle. The rectangle in Fig. 3.5 represents the *Robot* object class. The state net inside the rectangle represents the expected states for each object of the *Robot* object class.

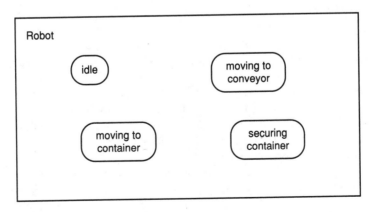

Figure 3.5   State net representation of the states of a manufacturing robot.

With respect to a particular object, a state is either on or off at any instant in time. We say that a state is *on* when the object is currently in the state. A state is *off* when the object is not in the state. In the Green-Grow Seed Company, for example, there may be several robots. Perhaps one is in the *securing container* state of Fig. 3.5, and all the others are in the *idle* or *moving to conveyor* state. For a robot in the *securing container* state, we say that the *securing container* state is on, and all other states are off. Similarly, for a robot in the *moving to conveyor* state, this state is on, and all other states are off.

3.4.2 Transition Modeling

In a state net, we choose to represent a trigger and an action description for a transition in a rectangle divided into two sections as Fig. 3.6 shows. The top section contains a trigger description. A trigger gives the conditions that, when met, may cause the transition to fire. The bottom section describes the action that takes place during the transition between states. The action may be composed of one or more individual operations. Each individual operation comprising the action may or may not be executed each time the transition takes place. Therefore, each action may conceptually have multiple threads of execution. However, a transition is not complete until all threads of execution within its specified action are complete.

Sometimes we need to reference an individual transition. To provide an easily referenced transition identifier, we place a number or other identifier in square brackets above the upper left corner of a transition as Fig. 3.6 shows. The identifier is optional, and we often omit it when there is no need to reference the transition.

We illustrate our notation for transitions in Fig. 3.7, which shows part of a state net for a manufacturing robot. The state net defines the transition between the states *moving to container* and *securing container*. The trigger is in the upper part of the transition rectangle, and the action description is in the lower part. The *1* within square brackets is the transition identifier.

In Fig. 3.7 the at (@) symbol preceding the trigger description designates that the trigger is based upon an event. We read the at symbol as "at," "when," or "upon." The full trigger for the transition in Fig. 3.7 may, therefore, be read as *when arrives at container station*.

The transition represented in Fig. 3.7 involves three ordered steps: (1) leaving the *moving to container* state when the event *arrives at part station* occurs, (2) performing the actions defined for the transition, and (3) entering the state *securing container*. With respect to the transition in Fig. 3.7, we call the *moving to container* state a *prior state*, and we call the *securing container* state a *subsequent state*. When the robot is in the *moving to container* state, which is the prior state of the transition, we say that the transition is *enabled*. Triggers cannot cause transitions to fire unless the transitions are enabled. After completing the transition, the robot enters the *securing container* state, which is the subsequent state of the transition.

Figure 3.8 presents a time line showing ordered occurrences for the state net in Fig. 3.7. The time line begins with the robot in the *moving to container* state. At time t_1 the robot arrives at the container station. The robot's arrival at the station triggers the transition. Because of the mechanics of the robot in its current state, it does not respond

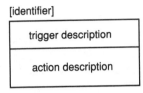

[identifier]

trigger description
action description

Figure 3.6   Representation of a transition's trigger and action description.

to the trigger and leave the *moving to container* state until time t_2. Figure 3.8 depicts this delay in response time by the gray area between time t_1 and t_2. Once the robot is out of the *moving to container* state at time t_2, the action of the transition takes place. The time between t_2 and t_3 is the total time required to complete the action and enter the *securing container* state. The time between t_1 and t_3 is the total time required for the overall transition after the transition is triggered.

During the time the transition action is being performed, t_2 to t_3, the robot is neither in the *moving to container* state nor in the *securing container* state. Instead, it is in a "state" of transition between the two states.

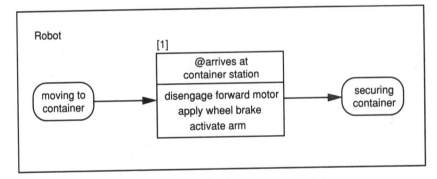

Figure 3.7   A partial state net for a manufacturing robot.

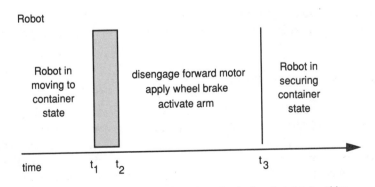

Figure 3.8   Time line showing time dependencies for robot state transition.

3.4.3 Concurrency

In OSA each individual object behaves concurrently. We refer to this concurrent behavior among objects as interobject concurrency. A robot, for example, might be moving to a container at the same time a packager is filling an order. Also, a packager may be in the transition between the states *cleaning up packing room* and *filling order* at the same time a different packager is in the transition between the states *filling order* and *addressing package*.

Furthermore, since each state net is a template describing the potential behavior of object instances, one or more objects may exhibit the exact same behavior simultaneously with other system objects. For example, two robots may both be moving toward seed containers at the same time.

OSA behavior models support interobject concurrency by associating a different state-net instance with every object. A state-net instance is an object's private copy of a state net. Although templates for a state net for two different objects in the same object class are identical, the two objects behave independently. Each may be in the same or different states and transitions, without regard to the other. Unless explicitly constrained by an analyst, the states and transitions of an object are independent of the states and transitions of any other object. Thus, objects in OSA naturally have concurrent behavior.

Besides interobject concurrency, objects in OSA may also exhibit intraobject concurrency. Intraobject concurrency allows an individual object to exhibit concurrent states or actions. A person, for example, may be talking on the phone while taking notes. A copy machine can copy and staple at the same time.

OSA models intraobject concurrency by a state net that allows an object to be in more than one state at the same time or combinations of states and transitions at the same time. Thus, in an object's private copy, an individual object may simultaneously be in more than one state or transition.

An analyst may express limitations to both interobject concurrency and intraobject concurrency in a state-net diagram through carefully chosen triggers, state definitions, and general constraints noted on the diagram.

Concurrency is an important issue in OSA. We have briefly introduced it here, but we return to it several more times as we continue our discussion of object-behavior models. In particular, we will be interested in how individual objects enter and leave concurrent states for intraobject concurrency. Before we can address these issues, however, we must first explain some additional concepts about object-behavior models.

3.5 Trigger Conditions and Events

As discussed earlier, a trigger causes an enabled transition to fire for an object when specified system events take place or certain system conditions are satisfied. These events and conditions may be internal or external to the object, and any combination of events and conditions may constitute a trigger. In general, a *trigger* is a boolean expression over expected events and conditions yielding a *true* or *false* result. In this section, we discuss condition-based triggers and event-based triggers in more detail.

3.5.1 Condition-Based Triggers

A *condition* is a logical statement about the current state of an object, the current state of the system environment, the existence or absence of an object, or the existence or absence of relationships among objects. Any logical statement, formal or informal, may be used as a trigger condition. The level of formality for a trigger condition's logical statement is determined by the requirements of the modeling task at hand.

Condition-based triggers cause an enabled transition to fire when their logical statement is true. Figure 3.9 shows an example. Here, a packager for the Green-Grow Seed Company may be in the *clean up packaging room* state. When the logical statement *packaging room is clean* holds, the transition fires. As a result, the packager enters the *idle* state.

Figure 3.10 shows the timing considerations for condition-based triggers. Figure 3.10(a) shows that a transition enabled by its prior states can fire as soon as its condition-based trigger is satisfied. Figure 3.10(b) shows that if a condition-based trigger is satisfied, the transition can fire as soon as it becomes enabled. Figure 3.10(c) shows a transition that will not take place because the transition is not enabled during the time its condition is satisfied.

3.5.2 Event-Based Triggers

An *event* is any change within a system. Examples include the creation or deletion of an object, a change in a relationship set, a change of state for an object, starting or stopping an activity, and the reception of a command or message. Any change that can be detected may be modeled as an event. Thus, when an object responds to an event, an object is responding to some detectable system change.

As with conditions, events are detected in state nets using triggers. We call the component of a trigger that detects event occurrences an *event monitor*. Event monitors are conceptual devices that "observe" a certain type of event in a system. An event monitor name associates the trigger with the type of event it monitors. For example, the event monitor named *@arrives at container station* for the robot example in Fig. 3.7 detects new instances of "arrives at container station" events.

Figure 3.11 illustrates the timing considerations for event-based triggers. The distinction between events and conditions is that an event triggers an enabled transition

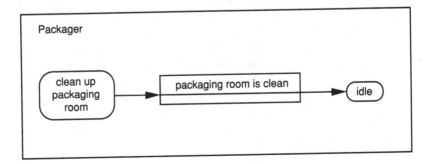

Figure 3.9   A condition-based trigger in a transition.

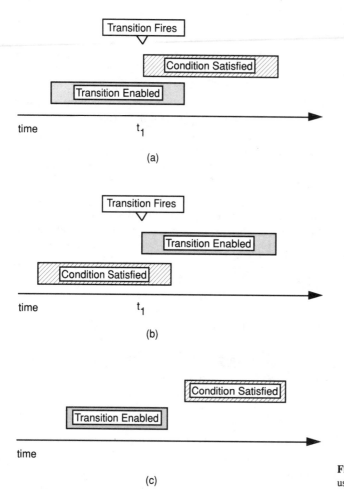

(a)

(b)

(c)

Figure 3.10   Timing considerations when using conditions as triggers.

only at the instant the event occurs, whereas a condition triggers a transition during the entire time the condition is satisfied. Figure 3.11(a) shows a transition that is enabled from time t_1 to time t_3 and an event monitored by the trigger of the transition that occurs at time t_2. Since the event occurs during the time the transition is enabled, the event can trigger the transition. Figure 3.11(b) shows a transition that is enabled from time t_2 to time t_3 and monitored events that occur at times t_1 and t_4. Since the events do not take place during the time the transition is enabled, the events cannot trigger the transition. Thus a trigger based on an event monitor can only cause a transition to fire if the transition is enabled at the time the event occurs. If the transition is not enabled, the event will go unnoticed by the object.

3.5.3 Events as Objects

In OSA we may model any event as an object, and we may group similar events together in object classes. Like any other object, event objects may participate in relationships.

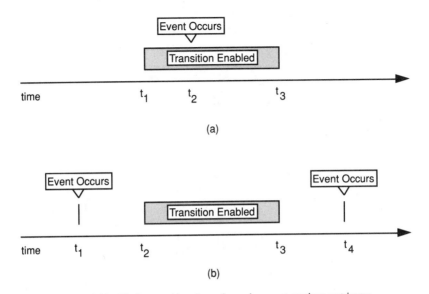

(a)

(b)

Figure 3.11 Timing considerations when using event monitors as triggers.

Modeling events as objects provides us with a way to record events and record information about events.

Figure 3.12 shows an example. The object class *Arrives At Container Station Event* contains instances of *arrives at container station* events. It also shows that we can have relationship sets for recording information about events. In Fig. 3.12 we have the relationship set "*Arrives At Container Station Event marked arrival of Robot*" which records instances of robots arriving at the container station. We also have the relationship set "*Arrives At Container Station Event occurred at Arrival Time*" which records the time of arrival for the event.

One interesting way we may wish to use recorded events and information recorded about events is in triggers. Rather than use an event monitor, we may use a condition based on a recorded past event. Figure 3.13(a) shows an example of a trigger based on information recorded about an event in the ORM in Fig. 3.13(b). The state net declares that a packager returning to the *idle* state leaves work if *today's 5:00 p.m. whistle has blown*. If the quitting whistle had been modeled as an event-based trigger, a packager returning to the *idle* state shortly after the whistle had blown would not go home since the event would have already taken place. Recording historical events as objects in an ORM is not required, but may be useful for clarifying the meaning of triggers based on historical events.

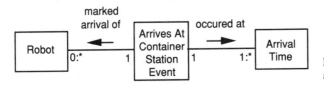

Figure 3.12 Events modeled as objects in an ORM diagram.

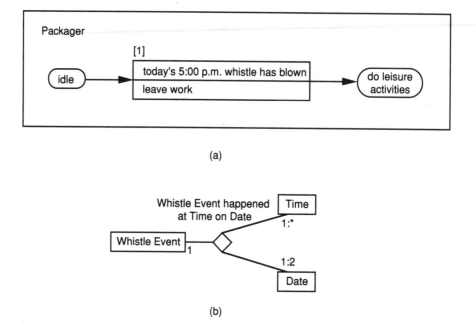

(a)

(b)

Figure 3.13   A trigger based on a historical event.

3.5.4 Compound Triggers

Conditions included in boolean expressions with event monitors form compound
triggers. Figure 3.14, for example, shows a compound trigger for a packager of the
Green-Grow Seed Company. The condition is a conjunction of an event monitor,
@manager enters the packaging room, and a condition, *packager feels guilty about
being idle.* In this example an idle packager only begins cleaning up the packaging
room if the manager enters the room and the packager feels guilty about being idle.
When a packager is idle and, either the manager does not enter the room or the packager
does not feel guilty about being idle, the transition does not fire.

We use the conjunction symbol (∧) to denote the *ANDing* of boolean expressions.
We also use the disjunction symbol (∨) to denote the *ORing* of boolean expressions.
However, any equivalent notation more suitable to the problem domain and audience is
allowable.

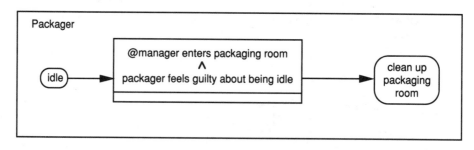

Figure 3.14   A compound trigger.

Thus, triggers may be any combination of ANDed and ORed event monitors and conditions. However, a conjunction of event monitors is suspect because event occurrence is instantaneous. It is unlikely that any two system events occur at exactly the same time.

The timing diagrams shown previously for condition-based and event-based triggers should be carefully considered when constructing compound triggers. Careful consideration should also be given to ensure that a compound trigger matches the real-world behavior of the system being studied and is not being used simply to shortcut the behavior modeling process.

3.6 State-Net Configurations

In the previous sections, we have explained and illustrated many state-net features. In this section, we provide additional detail about state nets. Our objective here is to more precisely describe and illustrate possible state-net configurations. To support this objective, we will often use state-net configurations with generic names for triggers, actions, and states. For example, we will use state names of S_1, S_2, and S_3 and trigger and action names a, b, c and d. These kind of examples are meant as state-net "templates" that the analyst may apply when discovering similar behavior occurrences in a system.

3.6.1 Subsequent States

Subsequent states are states that follow a transition. We choose to designate subsequent states for a transition in a state net by an arrow whose tail connects to a transition rectangle and whose head(s) connect to the state symbol for the subsequent state(s). Figure 3.15(a) shows an example where there is one state on the head of a directed arrow, and Fig. 3.15(b) shows an example where there are two heads and thus two states on the heads of a split directed arrow.

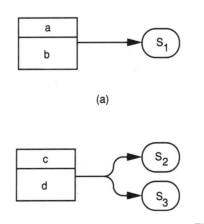

(a)

(b)

Figure 3.15 Subsequent states turned on at the completion of a transition action.

A multiheaded arrow leaving a transition and entering several state symbols designates that each state at which the directed arrow points is turned on at the completion of the transition. Thus, after completion of transition action b in Fig. 3.15(a), state S_1 is turned on, and after completion of transition action d in Fig. 3.15(b), both states S_2 and S_3 are turned on.

Figure 3.16 shows a real-world example of an object simultaneously entering and exhibiting two states at the same time. The state net shows the behavior of a copy machine after a paper jam has been cleared. The state net declares that a copy machine leaves the *ready after paper jam* state when the user signals for copying to continue. During the transition, the copy machine determines which sheets are unusable and what its restart page number should be. When these actions are complete, the copy machine enters both the *discarding unusable sheets* state and the *resetting document feeder* state. The copy machine then performs actions in these states simultaneously.

Besides single-headed and multiheaded out-arrows, a transition may also have multiple out-arrows pointing to subsequent states. Figure 3.17 shows an example with two directed out-arrows. One directed out-arrow points to state S_1 and the other points both to state S_2 and to state S_3. If a transition has more than one out-arrow pointing to subsequent states, there is a choice of which out-arrow to use. One and only one out-arrow will be chosen. The selection of which arrow to use is nondeterministic unless there is an applicable constraint on the state-net diagram that influences the choice.

Figure 3.18 shows an example of multiple out-arrows in the Green-Grow Seed Company example. In Fig. 3.18(a), after a packager addresses a package and places it in a bin, the packager decides whether to become idle or to clean up the packaging room.

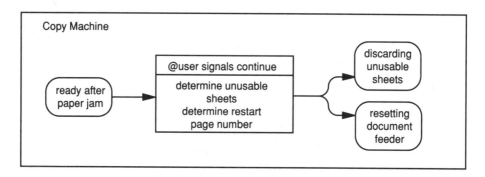

Figure 3.16   Modeling intraobject concurrency using multiple subsequent states.

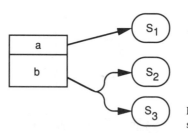

Figure 3.17   Multiple out-arrows from a state transition.

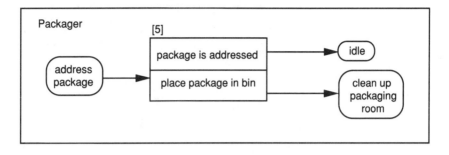

(a)

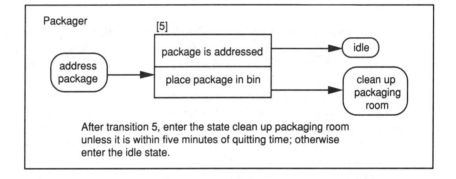

(b)

Figure 3.18   Packager state net with multiple out-arrows.

In this case, since no general constraints influence the decision, a packager may arbitrarily decide what to do.

Figure 3.18(b) shows an example where the decision on which state to enter following the transition is captured in a general constraint. The following general constraint allows us to declare the precise behavior path to take following the completion of transition 5 for the packager.

> *After transition 5, enter the state clean up packaging room unless it is within 5 minutes of quitting time; otherwise enter the idle state.*

3.6.2 Prior States

Prior states are states that precede a transition. We choose to designate prior states for a transition in a state net by an arrow whose head points to a transition rectangle and whose tail(s) connect to state symbols of prior states. Figure 3.19(a) shows an example where there is one state on the tail of a directed arrow, and Fig. 3.19(b) shows an example where there are two states on the tails of a directed arrow.

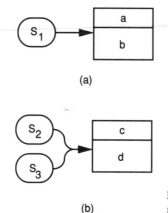

(a)

Figure 3.19 Transitions enabled by prior
(b)   states.

When all states at the tail of an in-arrow for a transition are on, the transition at the head of the arrow is *enabled*. In Fig. 3.19(a), for example, when state S_1 is on, the transition is enabled. To enable the transition in Fig. 3.19(b), both state S_2 and state S_3 must be on.

Figure 3.20 shows a realistic example using an in-arrow with two states at the tail. As our earlier example in Fig. 3.16 shows, when a copy machine jams and a user signals for continuation, the copy machine enters both the *discarding unusable sheets* state and the *resetting document feeder* state. When the copier is in these two states, then when the machine is ready following all jam recovery activities, the state net in Fig. 3.20 shows that the transition fires and the copier resumes copying and stapling.

In addition to single-tail or multiple-tail transition in-arrows, there may be multiple in-arrows connected to a transition. Figure 3.21 shows an example. When more than one directed arrow flows into a transition rectangle, the transition is enabled when the states at the tails of at least one of the directed arrows are all on. Thus, the transition in Fig. 3.21 is enabled if state S_1 is on or if both states S_2 and S_3 are on or if all three states are on.

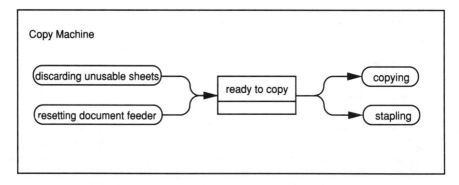

Figure 3.20 Copy machine state net with multiple-tail in-arrow.

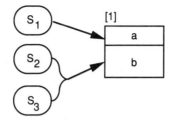

Figure 3.21   A transition enabled by multiple prior states.

When all three states in Fig. 3.21 are on, there is a choice of which enabling directed arrow to use. One arrow is chosen either at random or according to a constraint provided as part of the state net. For example, if we have the general constraint

The enabling states S_2 and S_3 have priority
over the enabling state S_1 for transition 1

then when all three states are on, the directed in-arrow connecting to states S_2 and S_3 is selected.

When a trigger fires for a transition, all states at the tails of the selected in-arrow are turned off. If there are other in-arrows, states at their tails are not affected by firing the transition. They remain on or off as they were.

A prior state that enables more than one transition is another configuration of interest. Figure 3.22(a) shows an example. When state S_1 is on, both transitions *1* and *2* are enabled. If neither trigger holds, neither transition fires. If only one of the two triggers hold, the transition whose trigger holds fires. If both triggers hold, only one of the two transitions fires. In the latter case, unless otherwise directed by general constraints, the choice of which transition fires is nondeterministic. When the chosen transition fires, state S_1 turns off and thus disables the other transition.

Figure 3.22(b) shows a more complex example. Suppose both states S_1 and S_2 are on and the triggers of both transitions hold. Both transitions fire if transition *1* selects state S_1 as its enabling state. However, if transition *1* selects state S_2 and locks it for its own use before transition *2* locks it for its use, then only transition *1* fires. Since state S_2 will have been turned off before transition *1* releases it, transition *2* will have become disabled and will not fire.

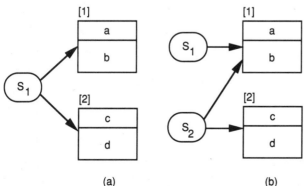

(a)   (b)   **Figure 3.22**   Two transitions in conflict.

Figure 3.23 shows an example of a state enabling several states simultaneously. When a packager is in the *idle* state, transitions *1, 4, 6, 8,* and *9* are all enabled. If there were no applicable constraints and if all triggers were to hold, a packager in the *idle* state would have to choose which transition to follow. Since there is an added constraint that imposes a priority order on the transitions, however, there is no choice. Transition *4* has top priority followed in order by transitions *1, 6, 9* and *8.*

3.6.3 Initial Transitions

Initial transitions activate initial states, which are those states exhibited by an object when it comes into existence in a system. Initial transitions have no prior states and are always enabled. Therefore, an initial transition fires whenever its trigger is satisfied. Every complete state net must have an initial transition.

An object creation event occurs for each object in a system as it comes into existence in the model. An event monitor that we often use for detecting the creation of a new object in a system is *@create*. Many initial transitions use this event monitor, or its equivalent, as a trigger. Equivalent event monitors serving the same purpose, for example, might be *@birth* for people coming into existence or *@hire* for an employee coming into existence as far as the hiring company is concerned.

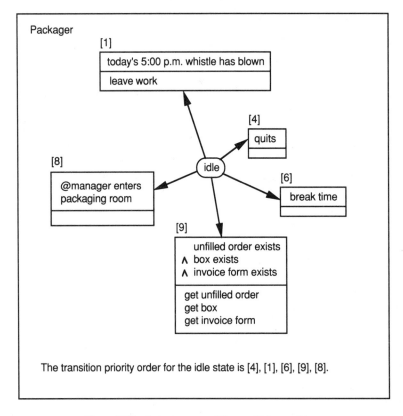

Figure 3.23   A single state enabling multiple transitions.

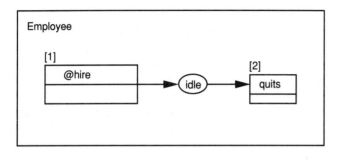

(a)

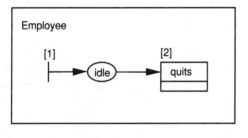

Figure 3.24 shows two state nets that initialize an employee object for the Green-
Grow Seed Company. Both initial transitions describe the creation of an *Employee*
object as far as the model is concerned. The state net of both Figure 3.24(a) and (b)
show that *idle* is the initial state of *Employee*. Figure 3.24(a) shows the initial transition
as a rectangle whose trigger is *@hire* and whose set of transition actions is null. When
an initial transition's trigger is semantically equivalent to *@create* and it has no transition actions, we have a shorthand representation. Figure 3.24(b) shows our shorthand
notation for an initial transition which is a bar across the tail of a directed arrow into an
initial state. Use of the shorthand bar means that the trigger is *@create,* or its
equivalent, and that there are no initializing actions.

(b)

Figure 3.24   Initial and final state
transitions.

In an initial transition, an analyst may specify initializing actions. If such actions
are required, we would not use the shorthand notation. Instead, we would use the rectangular notation and place the initializing actions in the action-specification part of the
transition. Figure 3.25 shows an initial transition with an initializing action. As part of
an initial transition for hiring a new employee, the initializing action calls for the
employee to sign an employee identification card.

3.6.4 Final Transitions

Final transitions are transitions with no subsequent states. When a final transition fires,
the prior states of the final transition are turned off. If no other states are on for the

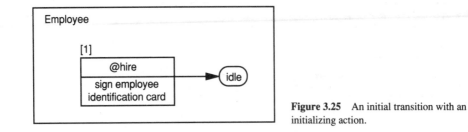

Figure 3.25   An initial transition with an
initializing action.

object, it ceases to exist as far as the model is concerned. In Fig. 3.24, the *idle* state for
an employee connects to a final transition whose trigger is *quits*. When the employee is
in the idle state and quits, the idle state is turned off and the employee no longer exhibits
any state in the system. Therefore, in the context of the Green-Grow Seed Company, an
employee is no longer part of the system model after the employee "quits." Final transitions are optional and are only required for objects expected to leave the system's
domain.

As with initial transitions there is a common event, *@destroy* or its equivalent, that
triggers an action causing an object to cease to exist in the system. Figure 3.26(a) shows
an example of a final transition with an *@destroy* trigger. When a weekly report is on
file, and the destroy event occurs, the transition takes place and the weekly report no
longer exists. Figure 3.26(b) shows an abbreviation for destroying a weekly report.

As with initial transitions, if there are actions in the final transition, we place them
in a rectangular transition box. Figure 3.27 provides an example. When an employee is
fired, the employee must return the employee identification card.

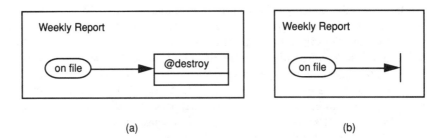

(a)   (b)

Figure 3.26   A shorthand notation for a final transition.

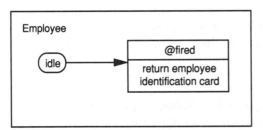

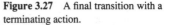
Figure 3.27   A final transition with a
terminating action.

3.6.5 Shorthand for Specifying Transitions

Many transitions have a single entry and a single exit. We have a shorthand notation for specifying these single-entry/single-exit transitions that is similar to traditional finite state machine notation. Figure 3.28 shows an example. Instead of placing a trigger *a* above a horizontal line in a rectangular transition box and a transition action *b* below the line in the box, an analyst may discard the box (but not the horizontal line) and directly connect the in-arrow and out-arrow as Fig. 3.28 shows.

The shorthand notation is particularly appropriate when there is no specified transition action. In this case, we also omit the horizontal line and provide only the trigger. Figure 3.29 shows an example. This state net declares that a packager in the *idle* state makes a transition to the *on break* state as soon as it is break time. No action is required to make the transition. Since there is no action for the transition, we only place the trigger description on the transition line between the state symbols.

In certain cases, we can further reduce the notation required for a trigger. A state transition may not require an event or condition other than the completion of an activity modeled by a prior state and may also not require an action. We may model these transitions by connecting a prior state to a subsequent state with an unlabeled arrow. Figure 3.30 shows an example. The state net declares that a clerk enters the *ready to work on orders* state as soon as the *on break* state is completed. In other words, after taking a break, a clerk is ready to work on orders.

Although it may now appear that we can directly connect a state to another state, this conclusion is false. Conceptually, there is still a transition in the diagram. The trigger for the transition is *finishes prior activities*, or its equivalent, and there is no action associated with the transition.

Besides simplifying transition specification, we also have a shorthand for eliminating some directed arrows. We use double-headed arrows when an object in some state

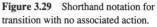

Figure 3.28   Shorthand state transition notation.

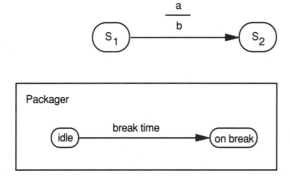

Figure 3.29   Shorthand notation for transition with no associated action.

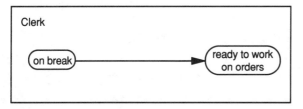

Figure 3.30   Shorthand notation for transition with an implied event and no action.

responds to a trigger and then, after performing any actions in the transition, returns to the same state.

Figure 3.31 shows an example with two double-headed arrows. A receptionist in the *ready to serve person* state may answer the phone or help the next person waiting for the receptionist. If the receptionist is in the *ready to serve person* state and the phone rings, the receptionist handles the call and then returns to the *ready to serve person* state. If a person in the lobby needs service, the receptionist enters the *serving person* state and when finished serving the person, returns to the *ready to serve person* state. If while the receptionist is in the *serving person* state the phone rings, the receptionist handles the phone call and then returns to the *serving person* state.

The general constraint on the diagram ensures that a receptionist finishes serving one person in the lobby before beginning to serve someone else in the lobby. This is accomplished by the *resume* clause in the constraint. If the constraint were omitted entirely, the receptionist could return to the *serving person* state, but serve a different individual than the one served prior to the phone call.

The double-headed arrow may be used in many ways. It is particularly useful for describing the behavior of what some might consider "stateless" objects. These are usually objects that have only one interesting state, their state of existence in the system. We can model interactions with this kind of object with double-headed arrows as transitions leaving from and returning to the single existence state.

A bank account, for example, may be considered as a stateless object. Figure 3.32 shows a partial state net for a single-state bank account object. The *static* state is the enabling state for the triggers in transition *2, 3, 4* and *5*. When the actions triggered by *@deposit, @withdraw,* or *@print balance* are complete, the bank account returns to the *static* state.

3.6.6 Remaining in an Enabling State When a Trigger Fires

Sometimes we wish to model an object entering a new state without leaving the current state. Figure 3.33 shows an example. In this example an executive can be in the state

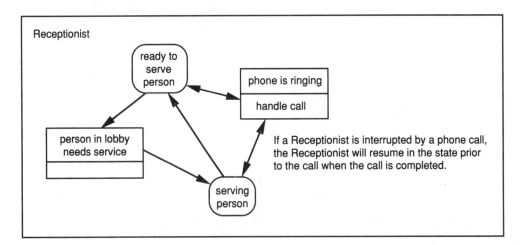

Figure 3.31   Use of double-headed arrows in a state net.

3.6.5 Shorthand for Specifying Transitions

Many transitions have a single entry and a single exit. We have a shorthand notation for specifying these single-entry/single-exit transitions that is similar to traditional finite state machine notation. Figure 3.28 shows an example. Instead of placing a trigger *a* above a horizontal line in a rectangular transition box and a transition action *b* below the line in the box, an analyst may discard the box (but not the horizontal line) and directly connect the in-arrow and out-arrow as Fig. 3.28 shows.

The shorthand notation is particularly appropriate when there is no specified transition action. In this case, we also omit the horizontal line and provide only the trigger. Figure 3.29 shows an example. This state net declares that a packager in the *idle* state makes a transition to the *on break* state as soon as it is break time. No action is required to make the transition. Since there is no action for the transition, we only place the trigger description on the transition line between the state symbols.

In certain cases, we can further reduce the notation required for a trigger. A state transition may not require an event or condition other than the completion of an activity modeled by a prior state and may also not require an action. We may model these transitions by connecting a prior state to a subsequent state with an unlabeled arrow. Figure 3.30 shows an example. The state net declares that a clerk enters the *ready to work on orders* state as soon as the *on break* state is completed. In other words, after taking a break, a clerk is ready to work on orders.

Although it may now appear that we can directly connect a state to another state, this conclusion is false. Conceptually, there is still a transition in the diagram. The trigger for the transition is *finishes prior activities*, or its equivalent, and there is no action associated with the transition.

Besides simplifying transition specification, we also have a shorthand for eliminating some directed arrows. We use double-headed arrows when an object in some state

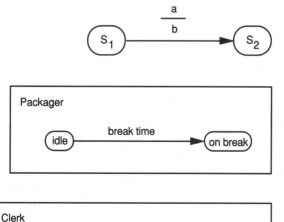

Figure 3.28 Shorthand state transition notation.

Figure 3.29 Shorthand notation for transition with no associated action.

Figure 3.30 Shorthand notation for transition with an implied event and no action.

responds to a trigger and then, after performing any actions in the transition, returns to the same state.

Figure 3.31 shows an example with two double-headed arrows. A receptionist in the *ready to serve person* state may answer the phone or help the next person waiting for the receptionist. If the receptionist is in the *ready to serve person* state and the phone rings, the receptionist handles the call and then returns to the *ready to serve person* state. If a person in the lobby needs service, the receptionist enters the *serving person* state and when finished serving the person, returns to the *ready to serve person* state. If while the receptionist is in the *serving person* state the phone rings, the receptionist handles the phone call and then returns to the *serving person* state.

The general constraint on the diagram ensures that a receptionist finishes serving one person in the lobby before beginning to serve someone else in the lobby. This is accomplished by the *resume* clause in the constraint. If the constraint were omitted entirely, the receptionist could return to the *serving person* state, but serve a different individual than the one served prior to the phone call.

The double-headed arrow may be used in many ways. It is particularly useful for describing the behavior of what some might consider "stateless" objects. These are usually objects that have only one interesting state, their state of existence in the system. We can model interactions with this kind of object with double-headed arrows as transitions leaving from and returning to the single existence state.

A bank account, for example, may be considered as a stateless object. Figure 3.32 shows a partial state net for a single-state bank account object. The *static* state is the enabling state for the triggers in transition *2, 3, 4* and *5*. When the actions triggered by *@deposit, @withdraw,* or *@print balance* are complete, the bank account returns to the *static* state.

3.6.6 Remaining in an Enabling State When a Trigger Fires

Sometimes we wish to model an object entering a new state without leaving the current state. Figure 3.33 shows an example. In this example an executive can be in the state

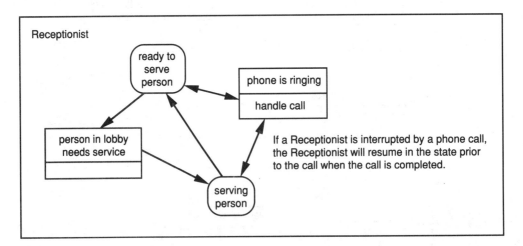

Figure 3.31   Use of double-headed arrows in a state net.

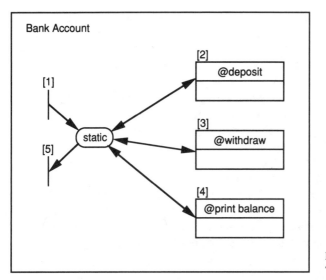

Figure 3.32   Modeling the behavior of "stateless" objects.

driving to work. We expect this state to be exited through transition *2* when the executive arrives at work. For the example we suppose that while driving to work, the thought suddenly comes that today is the birthday of the executive's spouse. When this happens, the executive picks up the car phone and dials a candy shop. After dialing the candy shop, the executive enters the state *ordering candy for spouse.* While this is happening, the executive is still in the state of *driving to work.* The executive therefore enters a new state, but does not leave the original state.

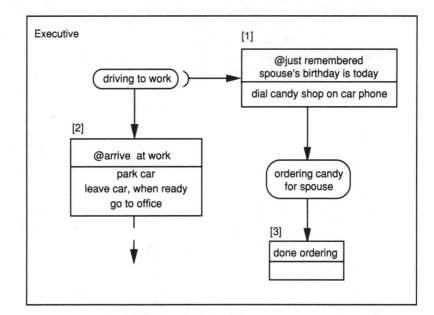

Figure 3.33   A state transition that does not cause the enabling state to turn off.

We have chosen to represent an enabling state that is not exited when an enabled transition fires by connecting the tail of the directed arrow pointing to the enabled transition to an arc near the enabling state. Figure 3.33 shows the tail of the in-arrow to transition *1* connected to an arc near the symbol for the *driving to work* state rather than to the state symbol itself.

Remaining in an enabling state when a trigger fires is a method for initiating intraobject concurrency. Whenever we have intraobject concurrency, we must properly terminate it. The state net in Fig. 3.33 shows one way to terminate intraobject concurrency. Transition *3* is a final transition. As discussed earlier, when an object exits a final transition it ceases to exist unless the object is concurrently in some other state or transition. In Fig. 3.33, an executive who has finished ordering candy will either still be in the *driving to work* state or will have arrived at work and will still be in transition *2*, which is not exited until the executive has parked the car and is ready to leave the car, which includes finishing talking on the car phone, if necessary. Thus, after transition *3* completes, the executive will be in only one state or transition and there will no longer be any intraobject concurrency.

3.7 Exceptions

An *exception* is a system event or condition that is not part of normal system behavior. We may, for example, consider a paper jam for a copy machine as an exception.

In our state-net notation we choose to show a state exception by drawing a bar across a transition arrow. The *Copy Machine* state net in Fig. 3.34 models the paper jam event as an exception by simply adding a bar across the in-arrow to the transition whose trigger identifies the exception condition.

The behavior specified by the state net is the same with or without the bar, but the bar allows an analyst to designate a particular path as an exception to the normal behavior. In Fig. 3.34, the normal exit from the *copying* and *stapling* states is to the *ready for next job* state when the copying and stapling are complete.

Exceptions may also occur during a state transition. We model a transition exception by drawing an arrow from the transition in which the exception may occur to a state that the object should enter following the exception. We also draw a bar through the

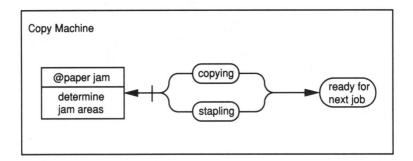

Figure 3.34   Modeling exceptions.

arrow near the transition rectangle to designate the connection as an exception and label the bar with a description of the exception.

Figure 3.35 shows a state net with a transition exception for the Green-Grow Seed Company. Transition *1* in Fig. 3.35 fires when a packager is idle. After the transition completes, the packager should fill an order by packaging items for the order in a box. The semantics of transition firing guarantee that when transition *1* fires, an unfilled order, a box, and an invoice form all exist. However, the box supply may fall to zero after the transition fires and before a packager can get an unfilled order, get a box, and get an invoice form. How might this occur? Suppose there are two packagers in the *idle* state, two unfilled orders, and two invoice forms, but only one box. The prior state would be on and the trigger for transition *1* would be true for both packagers. Thus both would leave the idle state and try to get an unfilled order, a box, and an invoice form. The first packager to take the last box will cause the supply to go to zero, causing an exception condition in the other packager's transition. The exception path in the state net declares that the packager failing to obtain a box is responsible for ordering more boxes. Once boxes are ordered, the packager should return to the *idle* state.

In Fig. 3.35 an exception could also have occurred if another packager had taken the last unfilled order or the last invoice form. Figure 3.36 shows how to handle all these exceptions.

In the state net of Figs. 3.35 and 3.36, the normal course of action for a packager who does not encounter an exception condition is to fill an order. We have said that when there are multiple out-arrows for a transition, the choice of paths is random in the absence of additional constraints. Exception arrows leaving a transition constitute one type of additional constraint. When an exception arises, it takes precedence over any normal exit.

Unlike state exceptions, which merely mark certain paths as being exceptions to normal behavior, transition exceptions add additional semantics to state nets. An object

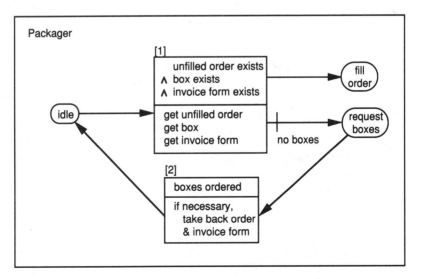

Figure 3.35   Exception during a transition of a packager.

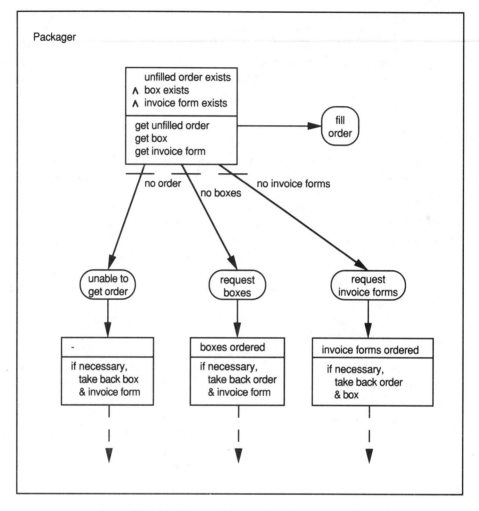

Figure 3.36   Modeling multiple exceptions for a state transition.

only follows an exception exit out of a transition if the exception condition associated with the exception exit holds. Furthermore, if an exception condition holds, the exception exit has precedence over normal exits. If the conditions of several exception exits from a transition hold, the choice of which exit to follow is arbitrary in the absence of added constraints.

3.8 Real-Time Constraints

When time is an important element of behavior, an analyst may add real-time constraints to state nets to specify timing requirements. We may apply real-time constraints to triggers, actions, states, and state-transition paths. We introduce real-time constraints by example, using the behavior of a packager.

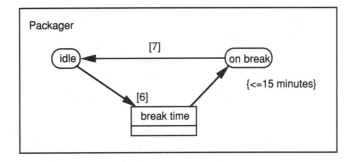

Figure 3.37   Real-time constraint applied to the duration of a state.

Figure 3.37 shows how to apply real-time constraints to a state. The state net in Fig. 3.37 describes a packager's behavior for a transition from the *idle* state to the *on break* state. The state net declares that a packager enters the *on break* state from the *idle* state when it is break time. The braces ({ }) near the *on break* state specify a real-time constraint for the duration of the state. The description within braces constrains the amount of time a packager may be in the *on break* state to a maximum of 15 minutes. Real-time constraint descriptions are free-format and therefore, may be tailored to match system requirements for formality and appropriate time units.

Figure 3.38 shows how to apply real-time constraints to the action part of a transition. The state net in Fig. 3.38 shows the behavior of a packager for the transition from the *request invoice forms* state to the *clean up packaging room* state. The braces in the bottom section of the transition rectangle enclose the real-time constraint, which constrains the duration of the transition's action. In this case, a packager must take 30 seconds or less to return an unpackaged order and unused box.

Figure 3.39 shows how to apply real-time constraints to the trigger part of a transition. The state net in Fig. 3.39 declares that a packager should move from the *idle* state to the *clean up packaging room* state whenever a manager enters the packaging room. The text within braces in the top section of the transition rectangle specifies a real-time constraint for the trigger. A trigger constraint specifies the maximum length of time allowed between the time the transition is triggered to the time the object leaves the prior states for the selected in-arrow. The state net in Fig. 3.39 declares that a packager in the

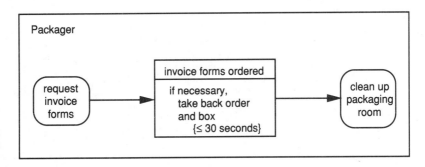

Figure 3.38   Real-time constraint applied to an action in a transition.

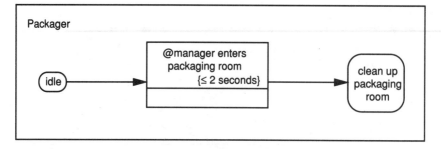

Figure 3.39   Real-time constraint applied to a trigger.

idle state must cease to be idle within 2 seconds after a manager enters the packaging room.

Figure 3.40 shows how to apply real-time constraints to an entire state transition. The state net in Fig. 3.40 shows that the transition between the *idle* state and the *fill order* state should take place when an unfilled order exists, a box exists, and an invoice form exists. The text within braces near the top of the transition contains the real-time constraint for the transition. In this case, the constraint declares that the transition should take 1 minute or less to complete.

If we specify real-time constraints for a transition and also for either its trigger and action part or both, we should make the real-time constraints consistent. It would not make much sense, for example, if we were to specify that the entire transition should take less time than the time allotted for the action alone. If real-time constraints for a transition happen to be inconsistent, each one applies individually as defined. Thus, for example, if the entire transition time is less than the trigger time alone, we simply require that both must be satisfied. The inconsistency is of little consequence since whenever the transition time is satisfied, the action time is also satisfied and could thus be omitted. For consistency's sake, the action time should either be omitted or one of the two constraints should be adjusted to be consistent.

The last type of real-time constraint is a constraint on a state-transition path. The state net in Fig. 3.41 describes the behavior for a packager during the process of packag-

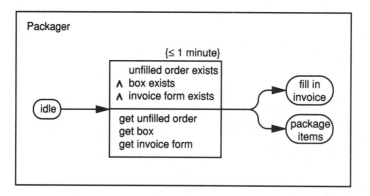

Figure 3.40   Real-time constraint applied to an entire transition.

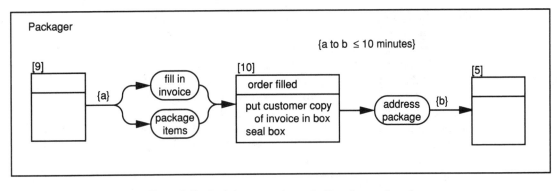

Figure 3.41   Real-time constraint used with state net path markers.

ing a customer order and addressing the package for shipment. The identifiers within braces ({a} and {b}) that are placed near directed arrows are called *path markers*. Path markers are used to mark begin and end times along state-transition paths. A path marker on a transition out-arrow marks the time that the subsequent states are entered. A path marker on a transition in-arrow marks the time that the prior states are exited. Therefore, the state net in Fig. 3.41 declares that a maximum of 10 minutes should elapse from time a packager begins filling a customer's order (marker {a}) to the time a packager has addressed the package containing the customer's order (marker {b}). Path marker identifier names are free-form, and the path duration description ({*a to b ≤ 10 minutes*} in Fig. 3.41) may be tailored to match the problem under study.

Most real-time constraints are added after an initial behavior model has been built. The state net in Fig. 3.42 shows real-time constraints added to the behavior model for

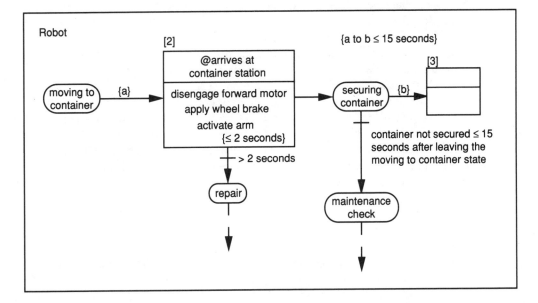

Figure 3.42   Real-time constraints and exceptions applied to a state net.

our manufacturing robot. Path markers have been added following the *moving to container* state and the *lifting container* state. The real-time constraint in braces declares that a robot should leave the *securing container* state 15 seconds or less from the time it leaves the *moving to container state*. There is also a constraint applied to the transition action, which declares that the action should complete in 2 seconds or less.

We may use exceptions to say what happens when objects fail to meet their real-time constraints. In Fig. 3.42, for example, if a robot fails to disengage its forward motor, apply its wheel brake, and activate its arm in 2 seconds or less, something is mechanically wrong. The robot, therefore, enters a *repair* state. Also in Fig. 3.42, if a robot fails to secure a container in 15 seconds or less after leaving the *moving to container* state, something may be wrong, and the robot enters a *maintenance check* state.

3.9 State Nets and Generalization-Specialization

As explained in Chap. 2, systems often provide an opportunity for abstracting similar classes of objects as a single, more general class—a process called *generalization*. Systems also provide the reverse opportunity to further refine the definition of members of a class by defining new specialized object classes—a process called *specialization*. These abstraction mechanisms also imply something about the behavior of objects in generalized or specialized classes. In this section we explain and illustrate the interrelationship between state nets for generalizations and specializations. We also explain how to ensure behavior consistency when generalization or specialization is used.

3.9.1 Developing State-Net Generalizations-Specializations

When we associate a state net with a particular object class, we declare that every member of the object class exhibits the behavior specified by the state net. From Chap. 2 we know that an object of a specialized class is also a member of a more general class. Therefore, an object of a specialized class must not only exhibit the behavior of the specialized class, but it must also exhibit all the behavior of the more general class. When we associate a state net with a specialized class, we must declare behavior that encompasses both the specialized class and the more general class. We must also be sure to account for all states and transitions declared for objects of the more general class.

We provide an example of state nets and class specialization using a state net for an employee of the Green-Grow Seed Company. The state net in Fig. 3.43 declares that an employee is initially in the *idle* state when the employee comes into existence by being hired. When it is time to leave work, the employee leaves the *idle* state and enters the *do leisure activities* state. The state net also declares that if it is time to go to work, an employee goes to work and enters the *idle* state. This state net is a template for the behavior of all employees of the company.

Figure 3.44 shows a partial state net for objects of the *Packager* object class which is a specialization of the *Employee* object class. The packager state net includes all the behavior of the employee state net. It also shows some additional behavior. When it is break time, a packager goes on break. If all employees of the Green-Grow Seed Company take official breaks, we could include this behavior in the employee state net. However, the manager, who is also an employee, does not take official breaks.

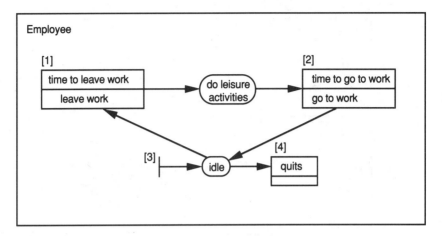

Figure 3.43 State net for an employee.

The corresponding state-net components for an employee in Fig. 3.43 and a packager in Fig. 3.44 are identical except for the trigger in transition *1*. The employee state net declares that an employee leaves when it is time to leave work, but the packager state net declares that a packager leaves when today's 5:00 P.M. whistle has blown. This is necessary because the Green-Grow Seed Company works packagers on an 8:30 A.M. to 5:00 P.M. shift. The trigger, *today's 5:00 p.m. whistle has blown*, for packagers is a stronger statement of the *time to leave work* trigger for employees. Therefore, the specialized behavior defined for packagers implies the more general behavior defined for employees.

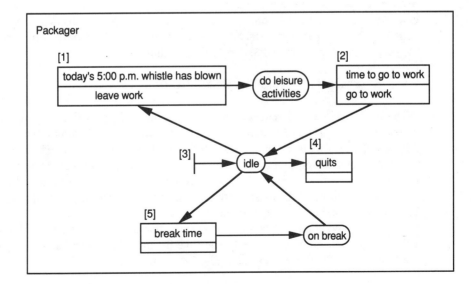

Figure 3.44 State net for a packager—a specialized employee.

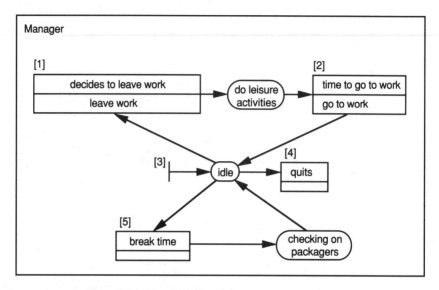

Figure 3.45 State net for a manager—a specialized employee.

The diagram in Fig. 3.45 shows a partial state net for the manager, who is also an employee. Again we see that the specialized behavior includes the generalized behavior. The trigger *decides to leave work* in transition *1* in Fig. 3.45 is the specialized, manager version of the more general *time to leave work* trigger for employees. In addition we see that a manager also checks on packagers, which is a behavior specific to managers, and is not a behavior of all employees.

When we specialize a state net, we can make triggers and actions more specific and we can add additional behavior. We can consider the state nets in Figs. 3.44 and 3.45 to have been created from the state net in Fig. 3.43 by specialization. We can also reverse the process, however, and generalize a group of state nets by identifying and extracting common behavior and, where necessary, by making triggers and actions more general. Thus, we can also consider the state net in Fig. 3.43 to have been created from the state nets in Figs. 3.44 and 3.45 by generalization.

3.9.2 Shorthand for State-Net Specialization

When the behavioral differences between a specialized object class and a more general class are minor or isolated with respect to affected states and transitions, we may wish to list only the new behavior instead of redrawing the entire specialized state net. The shorthand notation for simple state-net specializations accomplishes two things: (1) it reduces the potential for error since we only specify the changes, and (2) it produces a diagram where the differences between the general behavior and the specialized behavior are easier to see and understand.

Figure 3.46 shows an example of our shorthand notation by again presenting the state net specialization in Fig. 3.44. Here we list only the new behavior for packagers not found in the *Employee* state net in Fig. 3.43. The note

This state net specializes the Employee state net

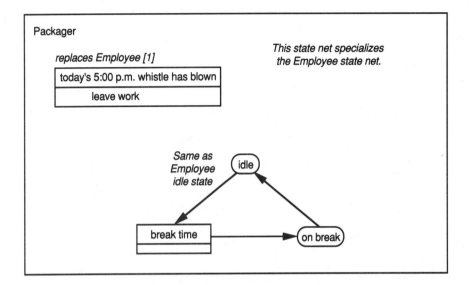

Figure 3.46   Shorthand specialization for packager state net.

tells us that the state net is a specialization and needs to be merged with the *Employee* state net to be read. Such notes are not required, but may be helpful in clarifying why the state net is incomplete and how to complete it. Other clarifying notes may also be needed. The note on top of the upper left transition tells us that this transition replaces transition *1* in the state net for *Employee*. The other note tells us that the *idle* state is the same as the *idle* state in the state net for *Employee*. Conventions such as using state name *Employee:idle* in place of state name *idle* and *Employee:1* within the brackets of the transition identifier could also be used and understood in this context.

When the transition of a more general class is replaced by another transition in a specialized state net, we are usually specializing a trigger or action in the transition. Here, we have specialized the trigger, as discussed previously. When the state of a more general class is referenced in a specialized state net, we are usually adding new connections to that state. Here, we connect a new state and transitions to show when a packager takes a break and what happens when the packager is done with a break.

If an object class's state net is represented using shorthand notation, we conceptually fill in all the details by merging it with the state nets of all generalizations above it in the generalization hierarchy. For complex state nets, many-level generalization hierarchies, or multiple inheritance, the merging of state nets can become conceptually difficult to manage. For these cases, we recommend using a standard state net.

3.10 A Sample State Net

We conclude this chapter by giving a state net in Fig. 3.47 for a packager working for the Green-Grow Seed Company. As always, an analyst is able to choose the level of abstraction for describing the behavior of an object. We have chosen to describe the behavior of a packager at the level of a typical job description for a person.

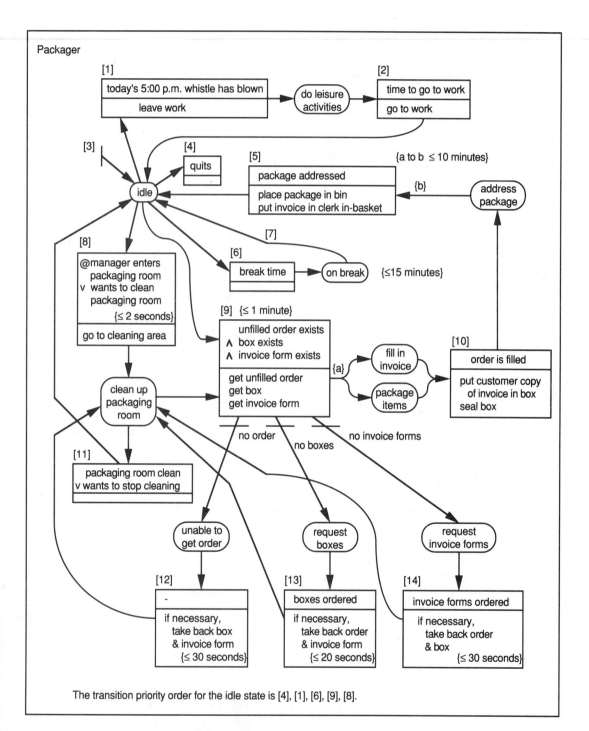

Figure 3.47   Full state net for a packager.

When a packager comes into existence (through transition *3*), the packager enters the idle state. There are several transitions out of the idle state, including transition *4* when a packager quits working for the company, transition *1* when it is time to go home, transition *6* when it is time for a break, transition *9* when an order should be filled, and transition *8* when its time to clean up the packaging room. If several or all the triggers of these transactions are simultaneously true, the general constraint at the bottom of the diagram in Fig. 3.47 gives a priority order for these transitions. Since transition *9* has priority over transition *8,* for example, filling an order has priority over cleaning up the packaging room. Quitting takes precedence over everything.

The main action of concern for a packager is packaging. Packaging is initiated in transition *9*. When an unfilled order is ready and there is a box and an invoice form, a packager begins filling the order by packaging requested items in the box and recording information about items placed in the box on the invoice form. After filling an order, a packager puts a copy of the invoice in the box and seals the box as a package to be sent. The packager then addresses the package, places it in the bin to be mailed, and puts a copy of the invoice form in an in-basket for a clerk. All this should happen in no more than 11 minutes since it should only take 1 minute to get set up to fill an order and 10 minutes to fill the order and address the package.

After finishing with an order in transition *5,* a packager enters the *idle* state. An idle packager goes to the cleaning area and begins cleaning the packaging room if either the manager enters the packaging room or if the packager decides to clean the packaging room. It takes less than 2 seconds once a decision is made to clean the packaging room to begin moving to the cleaning area. A packager who is cleaning the packaging room may be interrupted to fill another order. If there are none to fill, however, a packager cleaning the packaging room continues until the packager wants to stop cleaning or until the packaging room is clean.

The exceptions in the state net should be rare. The manager is supposed to make sure that there are boxes and invoice forms available and there should be plenty of orders for all the packages to fill. If, however, a packager competing with other packagers is left holding an item or two and cannot get the third to start filling an order, there is an exception. If the invoice form stack or box pile becomes empty, a packager caught in an exception path without an invoice form or box makes a request for the needed item. In any case, a packager returns items that could not be used and then begins cleaning the packaging room.

A packager goes home when the 5:00 P.M. whistle blows unless the packager is in the middle of filling an order, taking care of an exception, or cleaning the packaging room. Since it only takes a few minutes to perform any of these activities, the Green-Grow Seed Company asks its packagers to finish the activity before leaving. We are thus careful not to have packagers observe an event in transition *1*, but rather a condition indicating that an event (the blowing of today's 5:00 P.M. whistle) has occurred. After leaving work, a packager engages in leisure activities until it is time to return to work.

3.11 Bibliographic Notes

Many behavior models are not object-oriented. They were originally designed to model the behavior of an entire system under study. Several different types of system-wide behavior models have been developed. Davis describes many of them in his book [Davis 1990]. They include operational models, process-interaction models, finite state machines, decision-based models, stimulus-oriented models, and Petri nets.

The operational models describe behavior by defining how the system operates. Most describe the behavior using a programming paradigm [Jackson 1983, Orr 1981, Teichroew 1977, Warnier 1981]. PAISLey describes behavior using a functional programming model [Zave 1982].

Structured analysis techniques use process-interaction models (data flow diagrams) to model behavior [DeMarco 1979, Gane 1979, McMenamin 1984]. Other structured analysis techniques have extended the modeling power of data flow diagrams by incorporating finite state machines to model stimulus-based behavior [Hatley 1987, Ward 1985, Yourdon 1989]. To use finite state machines as a system-wide behavior model requires that the information be partitioned in some manner. Harel did this by extending finite state machines to allow for hierarchical decomposition and concurrency [Harel 1988].

The decision-based models incorporate decision trees and decision tables [Chvalovsky 1983, Moret 1982].

To organize the mass of detail contained in the behavior model of an entire system, some techniques have partitioned their information with respect to events or stimuli. One stimulus-oriented model, REVS, organizes behavior with respect to a single stimulus [Dyer 1977]. RLP organizes the information with respect to stimulus-response sequences [Davis 1979].

Many of the techniques include a concurrency modeling capability. Petri nets were developed to allow explicit modeling of concurrent actions [Peterson 1981].

Recently, behavior models have been proposed that organize information with respect to object classes. In OOA, Coad and Yourdon use a model that is conceptually similar to object-oriented languages [Coad 1991]. In OMT, Rumbaugh et al., use statecharts to describe the behavior of objects in an object class [Rumbaugh 1991].

Transitions in state nets differ from transitions of most other models. Unlike Petri nets and finite state machines, our transitions can take time to execute. We find that this feature supports a more accurate model of real-world object transitions. Our transitions are similar to high-level transitions as described by Peterson [Peterson 1981]. On the other hand, we can constrain our transitions so that they take no time to execute and thus we can restrict state nets to behave like finite state machines and Petri nets. The triggers of transitions also differ. In addition to firing when some stimulus or event occurs, they also fire when conditions hold. [Davis 1979, Dyer 1977, Hatley 1987, Ward 1985, Yourdon 1989]. In this sense they are similar to the guarded transitions of OMT where a transition only fires if the proper event occurs and the guard is true [Rumbaugh 1991].

Some of the simple behavior models do not support concurrency. Most of those that do, only support interobject concurrency [Coad 1991, Shlaer 1988]. We, like Rumbaugh et al., also support intraobject concurrency [Rumbaugh 1991].

In state nets, a transition may have more than one in-arrow and more than one out-arrow. The meaning of multiple in-arrows and out-arrows for a single transition is similar to the meaning of in-arrows and out-arrows for a single transition in a Petri net [Peterson 1981]. Although multiple in- and out-arrows are not supported by OMT, OMT does support single in- and out-arrows with multiple prior and subsequent states

[Rumbaugh 1991]. This is not true of Petri nets. Neither Petri nets nor OMT allows a prior state to remain unaffected when a transition fires. This expression of behavior is possible in state nets.

None of the current object-oriented analysis techniques supports exceptions. The technique developed by Myers allows exceptions for sequential logic, but his technique cannot be applied to concurrent environments [Meyer 1988].

Real-time constraints, similar to those used in states nets, can be found in many behavior models or their extensions [Coolahan 1983, Dyer 1977, Harel 1988, Taylor 1980, Wang 1988, Wasserman 1985, Zave 1988]. One unique capability of states nets is the use of real-time constraints to constrain triggers and actions.

Behavior generalization and specialization in object-oriented techniques, if found at all, is limited to method inheritance. In OSA, we also support state net generalization and specialization. In OMT the idea of generalization is discussed for states and events, but the generalization of individual object behaviors is not presented [Rumbaugh 1991].

3.12 Exercises

The structure of the automated teller machine (ATM) for the KWE banking system was described in the exercises section of Chap. 2. Here we give the behavior specification that will be used for the exercises in this and later chapters.

ATM.1.   When an ATM is off and the power is turned on, the ATM should: initialize itself, disable all user-interface (UI) module buttons, perform a self test, and establish the ATM network connection. After this, the ATM is ready to initiate customer interactions.

ATM.2.   When an ATM is ready to initiate customer interactions, the ATM should: enable card reader entry and prompt with "Welcome to KWE Bank. Enter Card." After this, the ATM should wait for the customer to enter a card.

ATM.3.   When an ATM is waiting for the customer to enter a card and the card is entered, the ATM should save the name and card number from the card and disable the card reader entry. The ATM should then validate the bank card.

ATM.4.   If the bank card is deemed valid, the ATM should: enable the numeric-entry group in the UI Module, enable the cancel button, and prompt with "Enter PIN or Cancel." The ATM should then wait for the customer to enter a personal identification number (PIN).

ATM.5.   When a PIN is entered, the ATM should: validate the PIN, prompt with a message welcoming the customer by name, display "Select Transaction." and button labels with transaction names according to instructions received by the central computer, and enable the transaction group buttons for which a label has been written. After this, the ATM should wait for the customer to push a button labeled with a transaction name.

ATM.6.   If the deposit button is selected, the ATM should: disable the transaction group, clear the button labels, prompt with "Enter Deposit Amount." and enable the numeric-entry group buttons of the UI Module. After this, the ATM should wait for the customer to enter the deposit amount.

ATM.7.   When a deposit amount is entered and is greater than zero, the ATM should disable the numeric-entry group and prompt with "Enter Deposit Envelope Now." and wait for the customer to enter the deposit envelope in the envelope slot.

ATM.8.　　When the deposit envelope is entered, the ATM should prompt with "Performing Requested Transaction. Please Wait." After this, the ATM should send the deposit transaction through the ATM network.

ATM.9.　　When the transaction completes successfully, the ATM should: prompt with "Transaction Successful. Printing Transaction Receipt.", print the transaction receipt, eject the bank card from the card reader, and prompt with "Remove Card from Card Reader." After this, the ATM should wait for the customer to remove the card.

ATM.10.　　When the card is removed, the ATM should be ready to initiate customer interactions once again.

ATM.11.　　When the ATM is waiting for a customer to push a transaction button and the withdraw button is selected, the ATM should: disable the transaction group, clear the button labels, prompt with "Enter Withdrawal Amount (in units of 20). Amount must be less than 200.", and enable the numeric-entry group buttons of the UI Module. The ATM should then wait to receive the withdrawal amount.

ATM.12.　　When the ATM is waiting for a withdrawal amount, and the amount is entered in units of 20, the ATM should disable the numeric-entry group and prompt with "Performing Requested Transaction. Please Wait." After this, the ATM should send the withdraw transaction through the ATM network.

ATM.13.　　When the transaction completes successfully, the ATM should prompt with "Transaction Successful. Printing Transaction Receipt." and simultaneously deliver the cash and print a receipt. When all the cash requested is in the cash drawer and the printing is complete, the ATM should prompt "Remove Cash from Cash Drawer." and wait until the customer removes all the cash.

ATM.14.　　When the cash is removed, the ATM should eject the bank card from the card reader and prompt with "Remove Bank Card from Card Reader." After this, the ATM should wait for the customer to remove the bank card.

3.1　Use a state net to model the behavior of the ATM (scenarios ATM.1 through ATM.14). Break this into manageable chunks as follows:

　a. Model the behavior scenarios ATM.1 through ATM.4.

　b. Model the behavior scenarios ATM.5 through ATM.10, and attach it to the state net of Exercise 3.1a.

　c. Model the behavior scenarios ATM.11 through ATM.14 and attach it to the state nets for Exercises 3.1a and b.

3.2　When the ATM is validating a PIN and the PIN is deemed invalid, the ATM should prompt with "PIN Is Invalid. Press Enter to Try Again." and recover by allowing the customer to enter another PIN. Using the transition in Fig. 3.48, show how to handle this exception.

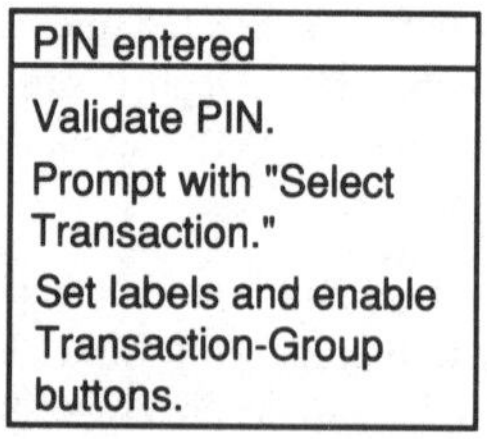

```
┌─────────────────────────┐
│ PIN entered             │
├─────────────────────────┤
│ Validate PIN.           │
│                         │
│ Prompt with "Select     │
│ Transaction."           │
│                         │
│ Set labels and enable   │
│ Transaction-Group       │
│ buttons.                │
└─────────────────────────┘
```

Figure 3.48　Transition for PIN entry exception.

3.3 When the ATM is waiting for a transaction deposit amount and the amount entered is zero, the ATM should prompt with "Deposit Must be Greater than Zero. Press Enter to Try Again." and recover by allowing the customer to enter another amount. Using a state labeled *waiting for deposit amount,* show how to handle this exception.

3.4 Suppose that a customer must remove an ejected bank card from the card reader within 60 seconds after the customer is prompted to remove the card. If the customer waits more than 60 seconds to remove a bank card, the ATM should pull the bank card into the machine and drop it into an internal bin for abandoned cards. The ATM should then prompt with "Bank Card Abandoned . . . Card Will Be Held Inside Machine Until Removal by Host Bank." for 15 seconds. After this, the ATM should be ready to initiate customer transactions once again. Using a state labeled *waiting for customer to remove card,* show how to model this behavior. Include a real-time constraint to model the 60 seconds.

3.5 Besides those exceptions described in Exercises 3.2, 3.3, and 3.4, identify several others (at least three) that would be necessary to handle erroneous inputs or conditions. Based on the behavior scenarios, specify the ATM state or transition action during which each exception would be detected.

3.6 A cash drawer for an ATM has a cash bin holding units of cash. When requested, the ATM puts units of cash equal to the withdraw amount in the cash drawer opening. For outdoor locations having a high risk of vandalism, there is a secure cash drawer with a security flap that seals the cash drawer opening when cash is not being delivered to the customer.

  a. Using the partial state net in Fig. 3.49 as a basis, model the behavior for a cash drawer.

  b. Model the behavior of secure cash drawers by showing the differences in behavior between a cash drawer and a secure cash drawer.

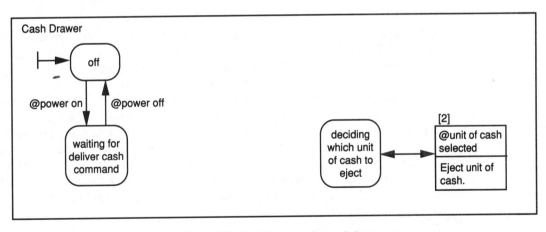

Figure 3.49 Partial state net for a cash drawer.

4

Managing Complexity
with Views

As models become large, they become increasingly complex and difficult to create, maintain, and display. The detail in large models usually makes them hard to read, understand, and alter. The detail also obscures the simplicity of the high-level concepts represented by the model. In this chapter we explain how to control and manage this complexity.

Our mechanism for reducing complexity in large models is abstraction. The main purpose of higher level abstractions is to represent fundamental system concepts, whereas lower level abstractions unfold supporting detail. To reduce complexity in OSA modeling, we provide multiple levels of abstraction for every type of construct—object classes, relationship sets, states, and transitions. In this chapter we show how to use abstraction as a means to view and manage large OSA models. In the chapters that follow we show how to use abstraction as a means to develop large models.

4.1 Views for Object-Relationship Models

High-level views of object concepts are common in our everyday conversation. We often refer to high-level concepts such as university, machine, and government when we are not interested in their make-up. We may also sometimes be interested in mid-level features without being interested in all low-level detail. We may, for example, refer to concepts such as department, college, faculty, students, and staff, which are all part of a university, but not mention any low-level details of these mid-level concepts.

High-level views for relationship-set concepts are also common. We often say that a person works in a city when, at a more-detailed level, we mean that the person works in an office located in the city. We also say that a person owns a house when we really mean either that the person is paying off a mortgage debt for the house or that the person (either singly or jointly with another person) has sole possession of the house. We talk about working in a city or owning a house at a high level of abstraction because we do not wish to become bogged down in all the detail that can be involved.

Without the ability to conceptualize high-level object classes and relationship sets, we would not be able to communicate easily. Low-level detail would constantly be in our way. In creating views for ORMs we will want the same facilities as we have in natural language to conceptualize and represent all-encompassing high-level views, intermediate mid-level views, and detailed low-level views. We also want to be able to switch easily from a high-level view to a low-level view and vice versa and be able to see the same concept from different points of view described at the same level of detail.

4.2 Object-Class Views

A *high-level object class* groups object classes, relationship sets, constraints, and notes into a single object class. For example, in the ORM for the Green-Grow Seed Company, which is duplicated in Fig. 4.1, we may wish to create a high-level object class by grouping together the object classes *Preferred Customer Group, Name* (of *Preferred Customer Group*), and *Discount Rate,* including the relationship sets that connect them, and the participation constraints on these connecting relationship sets. Figure 4.2(a) contains the part of Fig. 4.1 we are discussing and shows the high-level object class we are creating as a cloud. (We use the cloud in these examples because it is a convenient way to form a grouping of OSA modeling components that often do not conform to a simple polygonal border, such as a rectangle. Other methods of presenting these groupings may be chosen by the analyst.) We have chosen to call the new high-level object class *Discount Information* because the higher level conceptualization within the cloud can be thought of as an object class containing discount information for customers. To represent it as such and to discard its details, we implode the cloud and subsume its contents as a high-level object-class called *Discount Information* as Fig. 4.2(b) shows.

In Fig. 4.2(b), we no longer have the *Preferred Customer Group* object class, and we therefore can no longer have the associated *is member of* relationship set. So that we do not entirely lose this connection, we establish a link between *Customer* and *Discount Information* for the *is member of* relationship set. In our notation, we choose to represent this link as a dashed line as Fig. 4.2(b) shows.

A *dashed line* connecting two or more object classes denotes the existence of at least one relationship set connecting an object class within a high-level object class to an outside object class. In Fig. 4.2(a), for example, *Preferred Customer Group* is an object class within *Discount Information* that has a relationship set connecting it to the object class *Customer,* which is outside of *Discount Information*. If there are several such connecting relationship sets to the same outside object class, we represent them all as a single dashed line. The dashed line merely indicates that there is one or more connections, the details of which can only be seen in a lower level ORM diagram.

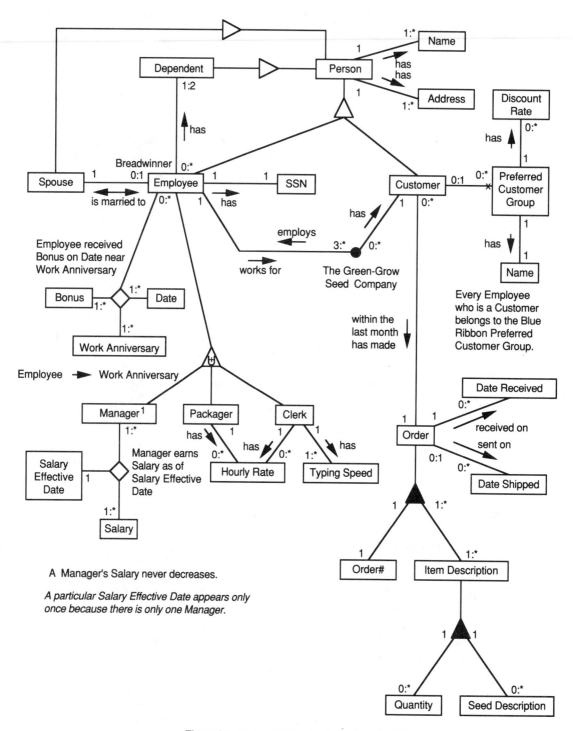

Figure 4.1   Basic ORM for the Green-Grow Seed Company.

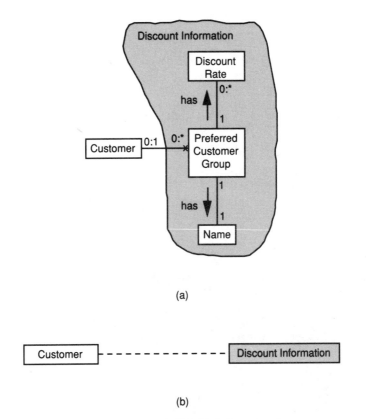

(a)

(b)

Figure 4.2   Creation of a high-level object-class view.

Referring again to Fig. 4.2(b), we see that *Discount Information* is shaded. Notationally, we have chosen shading as an indicator that a component is high-level. Other than the shading, there is no difference in appearance between a high-level object-class and an object class. Indeed, we need not even shade a high-level object-class when the intended use requires no knowledge of the existence of a more-detailed specification. Whether shaded or not, we may think of a high-level object-class as containing a set of similar objects in the same way we think of an object class as containing a set of similar objects.

4.2.1 Dominant and Independent Object-Class Views

Let us consider another group of object classes and relationship sets in Fig. 4.1 that constitute a high-level concept. At a high level of abstraction, it makes sense to think of an order as including the object classes *Order, Date Received, Date Shipped,* all the object classes in the part-of hierarchy under *Order,* including all the interrelating relationship sets, and all the participation constraints on the interrelating relationship sets. Figure 4.3(a) shows part of Fig. 4.1 and designates the suggested grouping with a cloud labeled *Order.* If we now implode the cloud, we obtain the diagram in Fig. 4.3(b).

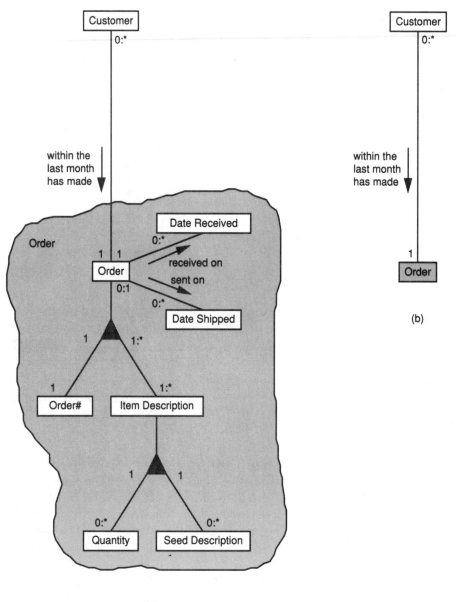

(a)

Figure 4.3 Creation of a dominant high-level object class.

In Fig. 4.3(b) we have subsumed the cloud under the name of one of the object classes included within the cloud. We thus have two ways in which we create a high-level object class, either by subsumption under the name of one of the object classes in the high-level class or by subsumption under a class name chosen independently. We refer to the former as subsumption by a dominant object class and the latter as subsumption by an independent object class concept. *Subsumption by a dominant object class* groups items under an existing object class. *Subsumption by an independent object class* groups items together that may be considered to represent some concept and then names the group with a concept identifier. We have subsumed *Order* in Fig. 4.3 under a dominant object class and *Discount Information* in Fig. 4.2 as an independent object class.

Besides the name, there is also another difference between subsumption under a dominant object class and subsumption as an independent object class. In Fig. 4.3(b), the relationship set *"Customer within the last month has made Order"* is not represented as a dashed line, but is fully represented as in Fig. 4.1 with its name and participation constraints. We are able to do this because the relationship set connects to the object class chosen as the dominant object class under which we subsume the high-level object class.

A relationship-set line that crosses the boundary of a group of object classes and relationship sets subsumed as an independent object class (rather than under a dominant object class) always becomes a conceptual link, and thus always becomes a dashed line in the view diagram. A relationship set line that crosses the boundary of a dominant high-level object class is also a conceptual link unless it connects on the inside solely to the dominant object class. A relationship set connected on the inside solely to the dominant object class retains its full details in the high-level view diagram. In Fig. 4.3(a), the only boundary-crossing relationship set connects the dominant object class *Order* to the outside object class *Customer,* and thus the full details of the relationship set are retained in the view diagram as Fig. 4.3(b) shows.

4.2.2 Nested Object-Class Views

When creating object-class views, existing high-level object classes behave as atomic object classes, and dashed-line relationship-set indicators behave as atomic relationship sets. (An *atomic* element in our model is one that does not contain any supporting detail and is, therefore, shown at its lowest level.) We may, for example, construct a dominant object class for *Customer* that includes *Customer, Discount Information,* and *Order* as Fig. 4.4 shows. In Fig. 4.4(a), *Discount Information* is the high-level object class in Fig. 4.2(b), and *Order* is the high-level object class in Fig. 4.3(b). In Fig. 4.4(b), *Customer* is an even higher level object class that includes customer-discount information and customer-order information within it.

By way of comparison, we construct an independent object-class view from the cloud in Fig. 4.4(a) rather than a dominant object-class view. We choose to call the high-level independent object class *Customer Information*. When we implode *Customer Information,* we obtain the diagram shown in Fig. 4.5, which is a little different from the diagram in Fig. 4.4(b). In Fig. 4.5 one specialization line in the *Person* specialization is dashed. In this context the dashed line indicates that an object class within *Customer Information* is a specialization of *Person*. It would be invalid for the dashed line to be solid, for then we would interpret *Customer Information* as a specialization of *Person* and thus as a subset of *Person,* which is incorrect.

4.2.3 Valid Object-Class Views

When we create a high-level object class, we wish to be sure that after its subsumption, we always have a valid ORM diagram. Other than imposing this requirement, we do not constrain choices for forming groups of object classes and relationship sets and for

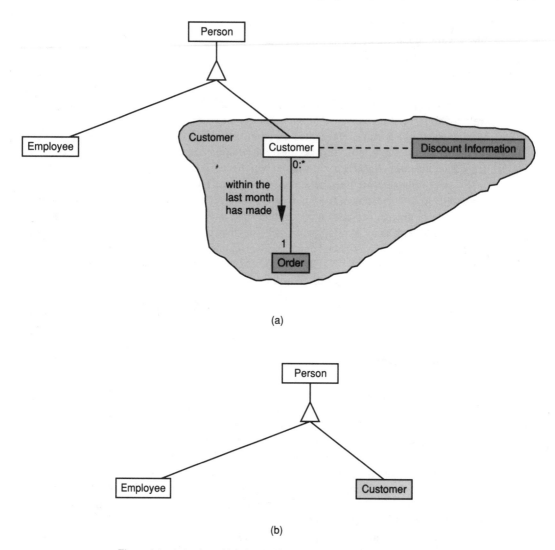

(a)

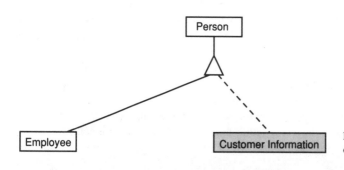

(b)

Figure 4.4   A dominant high-level object class with high-level object-class components.

Figure 4.5   An independent high-level object class with a hidden specialization.

designating dominant object classes. This probably allows greater freedom than is necessary, but since we cannot always be sure what will and will not make sense for a particular grouping, we prefer to provide the freedom and let the analyst decide.

One requirement that must hold in the creation of valid object-class views is that if we include a relationship set, then we must also include its participation constraints and all its associated object classes. It would not make sense in Fig. 4.2, for example, to omit the object class *Discount Rate* if we include the relationship set "*Preferred Customer Group has Discount Rate.*" Doing so, would leave a dangling relationship set in the new view.

4.3 Relationship-Set Views

A *high-level relationship set* groups object classes, relationship sets, constraints, and notes into a single relationship set. In Fig. 4.1, for example, we may wish to create a high-level relationship set showing *Discount Rate* connected directly to *Customer*. Figure 4.6 gives the part of Fig. 4.1 in which we are interested and shows how we create this view. The cloud in Fig. 4.6(a) includes *Preferred Customer Group,* its connected relationship sets and *Name*. The participation constraints for the new relationship set with its name are written next to the cloud. Figure 4.6(b) shows the new high-level relationship set after implosion. This view diagram nicely shows the discount rate for which a customer qualifies without the detailed explanation of why the customer qualifies for the rate.

In Fig. 4.6(b), we have chosen to shade the relationship-set diamond to indicate that the relationship set is high level. Although we have not normally used diamonds for binary relationship sets, we allow them. If the intended application has no need to be concerned with lower level relationship sets, we would not shade the diamond; and if the relationship set is binary and not shaded, we would omit the diamond.

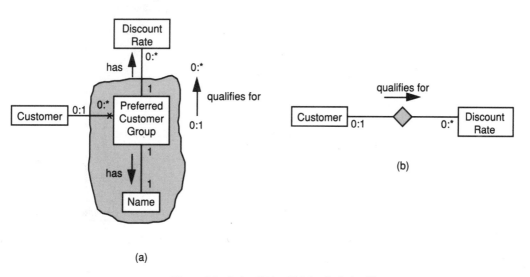

Figure 4.6   A simplifying high-level relationship set.

4.3.1 Participation Constraints in Relationship-Set Views

Because of our understanding of the diagrams in Fig. 4.6, we know that a single discount rate may be related to many customers. We also know from the participation constraints on the relationship sets connected to *Preferred Customer Group* in Fig. 4.6(a) that a customer may qualify for at most one discount rate.

Reviewing examples of high-level relationship-set creations such as the one in Fig. 4.6(a), could make one wonder about the possibility of automatic derivation of participation constraints. Unfortunately, even though automatic derivation is sometimes trivial, it is more often difficult and occasionally impossible. The impossibility arises when general constraints on the ORM apply to the derivation of high-level constraints. Since general constraints may not in a practical sense always be couched in formal terms, a specific algorithm cannot be constructed for automatic derivation.

Even though we cannot always guarantee consistency among low-level and high-level participation constraints, we do require that all stated participation constraints be satisfied. If an analyst is unable to give tighter bounds, $0:*$ is always permissible. If an analyst chooses to give tighter bounds than can be logically inferred from given constraints, these are accepted. Presumably, the analyst knows more than is implied by the given constraints.

4.3.2 Relationship-Set Views Constructed from Parallel Relationship Sets

Let us consider another example. Figure 4.7 is an ORM about persons and vehicles. A person may be leasing, may be buying, or may have the legal title to zero or more vehicles. A leased vehicle must be leased by some person and may also be jointly leased. Similarly, one or two people may be buying a vehicle or may have the legal title to a vehicle.

We wish to create a high-level view of the three relationship sets in Fig. 4.7(a) that simply states that a person owns a vehicle without explaining how a person owns a vehicle. The cloud in Fig. 4.7(a) indicates that we include the three relationship sets "*Person*

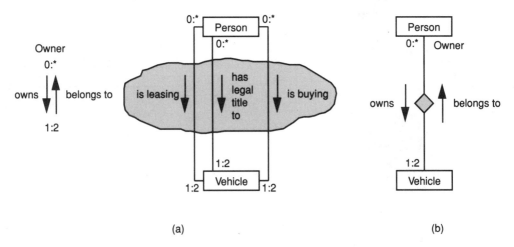

(a)   (b)

Figure 4.7   Parallel construction of a high-level relationship set.

is leasing Vehicle," "Person has legal title to Vehicle," and *"Person is buying Vehicle."* The information to the left of the cloud gives two different names for the high-level relationship set, the participation constraints, and a role for *Person* in the relationship set. After implosion, we have the high-level relationship set as shown in Fig. 4.7(b).

4.3.3 Valid Relationship-Set Views

We do not wish to restrict the choices of object classes, relationship sets, constraints, and notes that may constitute a high-level relationship set. However, we do wish to ensure that construction and implosion of each high-level relationship set is valid. One requirement that must hold to ensure validity is that if we include an object class in the creation of a high-level relationship set, then we must also include all its associated relationship sets. If we do not satisfy this requirement, we could create high-level ORMs with relationship sets connected to one another, which makes no sense.

We have not discussed subsumption of relationship sets under a dominant relationship set. Unlike high-level dominant object classes, we gain no advantage by introducing dominant relationship sets. If we wish to reuse a low-level relationship-set name at a high-level, we may do so, but the high-level name is independent of the low-level name. In Fig. 4.7, for example, if the relationship set *"Person has legal title to Vehicle"* had been *"Person owns Vehicle"* we could still have chosen to name the high-level relationship set *"Person owns Vehicle."*

4.3.4 A High-Level ORM for the Green-Grow Seed Company

To conclude our section on high-level views for ORMs, we present in Fig. 4.8 an example of the type of view diagram we might wish to create. Beginning with the diagram in Fig. 4.1, we can create the high-level ORM Fig. 4.8 as follows. We first create the high-level relationship set *"Customer qualifies for Discount Rate"* shown in Fig. 4.6(a) and implode it. We then create the dominant high-level object class *Order* shown in Fig. 4.3(a) and implode it. We next create a dominant high-level object class for *Person* that also includes the object classes *Name* and *Address* and their relationship-set connections to *Person*. Finally, we create the dominant high-level object class *Employee* that includes all the object classes in the generalization-specialization hierarchy rooted at *Employee* plus all associated object classes and relationship sets except *Person* and the singleton set for the seed company itself. Also included in the high-level *Employee* object class is the note *"A particular Salary Effective Date appears once and only once because there is only one Manager,"* and the constraints *"A Manager's Salary never decreases"* and *"Employee* $\longrightarrow$ *Work Anniversary."*

The result is the high-level view of the Green-Grow Seed Company in Fig. 4.8. The dashed line connecting *Person* and *Employee* in Fig. 4.8 shows that in addition to knowing that *Person* is a generalization of *Employee,* we also know that there is at least one other relationship between employees and persons. In this example there are two, one through *Spouse* and one through *Dependent.*

Observe that the constraint

Every Employee who is a customer belongs
to the Blue Ribbon Preferred Customer Group

from Fig. 4.1 is not in the high-level view in Fig. 4.8 nor is it included in any of the high-level object classes or relationship sets. The constraint restricts the relationship set *"Customer is member of Preferred Customer Group"* which is included within the high-level *qualifies for* relationship set. Unfortunately, the constraint also references the *Employee* object class which is neither an object class of, nor included in, the *qualifies for* relationship set. Thus, the constraint was (properly) omitted from the view in Fig. 4.6. Views that do not include all referenced object classes and relationship sets of a constraint should probably not include the constraint. Since the view in Fig. 4.8 does not include the *"Customer is member of Preferred Customer Group"* relationship set, we do not include it here either.

There can be good arguments for or against this practice depending on the audience, however, and since we generally wish to allow free expression whenever possible, we do allow an analyst to insist that a constraint be part of a view diagram even when the view does not include all referenced object classes and relationship sets. As a notational convention, we mark these constraints by underlining them with a dashed line. As with dashed connecting lines that indicate the existence of relationship sets at a lower level of detail, dashed underlines indicate that lower level details are needed to fully understand a constraint. We have a similar convention for notes. We underline a note with a dashed line if we want to display it even though it references an item not present in the current view.

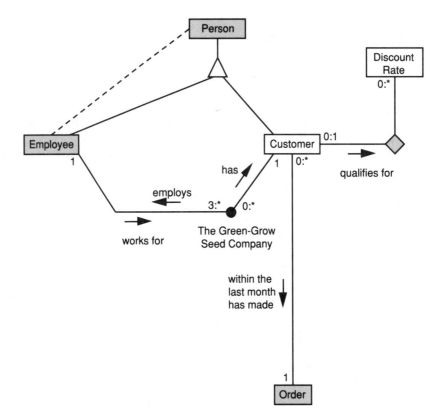

Figure 4.8   A high-level view of the basic ORM for the Green-Grow Seed Company.

4.4 High-Level Views for State Nets

In everyday conversation we commonly refer to objects as being in states and in transition between states. Much of the time, perhaps even most of the time, we discuss states and transitions at a high-level of abstraction. If we ask a child, "What are you doing?" the response may be, "Playing." If we want to know more detail about the state of the child, we may ask, "What are you playing?" The child may say, "I'm playing a game of chess with my friend." "Playing" is a high-level state that includes the lower level state "playing chess" within it. "Playing chess," also includes even lower level states within it and is, in this regard, a high-level state. A more precise description of the state of the child may be that the child is "taking a turn," or at even a lower level of description, the child may be "capturing the white queen." In our efforts to communicate with one another, we choose the level of abstraction for the state of an object that is most appropriate to the purpose at hand.

We may also consider transitions at various levels of abstraction. A worker may be in transition from work to home. A more detailed transition within "go home" might be "step onto the subway." "Step onto the subway" is a transition from the state "waiting for the subway" and "riding the subway," which are both states included within the higher level transition "go home."

As illustrated there are many levels at which we can communicate about states and transitions. We choose an appropriate level of abstraction and even change freely from level to level to best suit our needs in a conversation. We wish to provide these same facilities for state nets.

4.5 High-Level State Views

A *high-level state* groups lower level states, transitions, constraints, and notes into a single conceptual state. For example, in the state net for a packager, which we have copied into Fig. 4.9, we may wish to group states *package items* and *fill in invoice* into a higher level conceptual state called *fill order*. The diagram in Fig. 4.10(a) shows part of the packager state net with a cloud labeled *fill order,* which groups these states. This higher level conceptualization can be thought of as a state in which a packager is doing whatever is necessary to fill an order. Figure 4.10(b) shows this conceptualization as a high-level state. The shaded state symbol in Fig. 4.10(b), like the shaded symbols discussed earlier, indicates that the state is high level.

Similar to view creation for ORMs, when we create a new high-level state, any high-level states included in the view behave as if they were atomic states. Figure 4.11(a) shows an example. Here we are assuming that *fill order* is the high-level state created in Fig. 4.10. When we implode the cloud in Fig. 4.11(a), we obtain the high-level *packaging* state shown in Fig. 4.11(b).

4.5.1 Arrow Connections for State Views

The cloud in Fig. 4.12(a) shows the boundary of a high-level state, which includes all the states, transitions, and constraints for a packager who is at work, but is not packaging. After subsumption, Fig. 4.12(b) shows this new high-level *not packaging* state. In

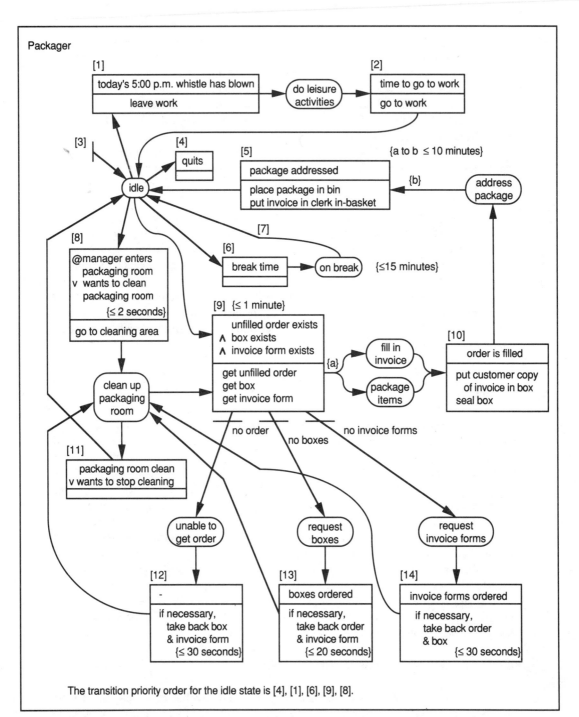

Figure 4.9 Packager state net.

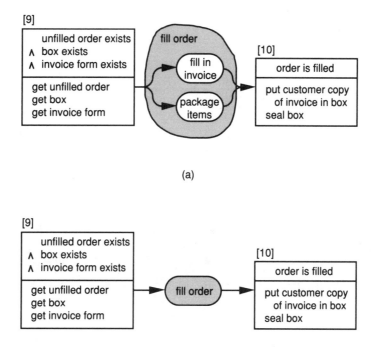

(a)

(b)

Figure 4.10   Construction of a high-level state.

addition to the usual subsumption of included states, transitions, and constraints, observe that there is also a reduction in the number of arrows incident on the high-level state. In Fig. 4.12(a) there are 10 arrows crossing the cloud boundary, but in Fig. 4.12(b) there are only 7 arrows incident on the new high-level state. Observe also that some of the arrows in Fig. 4.12(b) are dashed and that one arrow is an exception arrow.

Whenever two or more arrows have identical connecting tails and heads, only one is shown. In Fig. 4.12(a), for example, three exception arrows emanate from transition *9*. One goes to the *unable to get order* state, another goes to the *request boxes* state, and the other goes to the *request invoice forms* state. All of these states are subsumed in the high-level *not packaging* state. Thus, after subsumption, we would have three arrows with identical connecting tails and heads, all from transition *9* to *not packaging,* but we only show one in the high-level diagram.

Whenever several boundary-crossing exception arrows are coalesced into one because of having identical heads and tails, we are able to derive the exception condition as the disjunction of the exception conditions of the coalesced exception arrows. In Fig. 4.12(b) for example, the condition on the exception arrow from transition *9* to the *not packaging* state is a disjunction of the conditions of the exception arrows in Fig. 4.12(a).

The dashed arrows in Fig. 4.12(b) are introduced here for the first time. Unlike ordinary arrow connectors, which are solid and which are the only type of arrow connector we have discussed, *dashed arrow* connectors mean that an object may sometimes not traverse the arrow under the usual rules of states and transitions. To determine

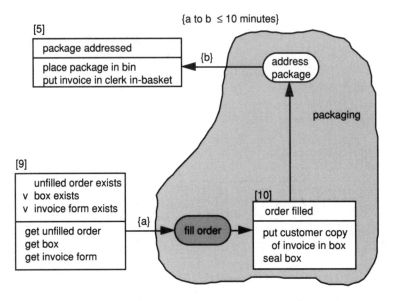

(a)

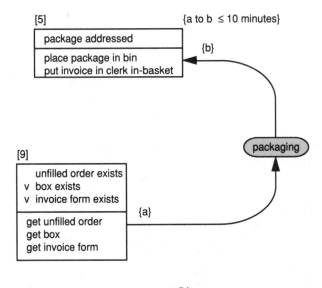

(b)

Figure 4.11   A high-level state with a high-level state component.

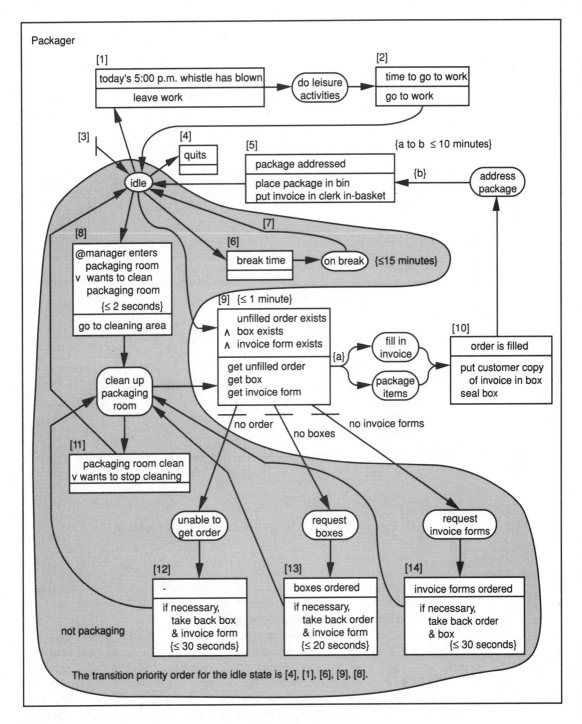

(a)

Figure 4.12 A high-level state with several simplifications.

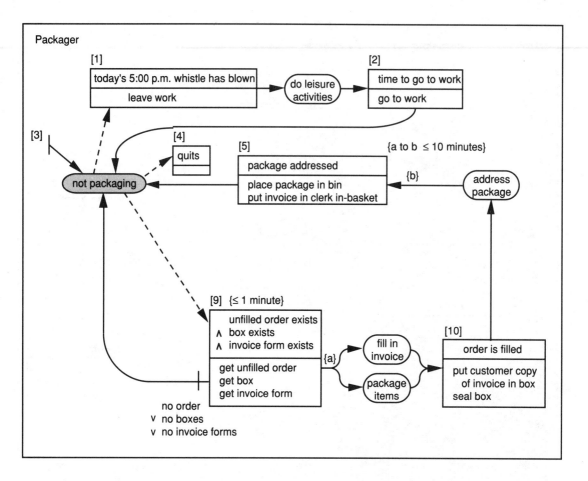

(b)

Figure 4.12 (Cont'd.)   A high-level state with several simplifications.

whether an object traverses a dashed arrow, we must consult a lower level state net. In this sense a dashed arrow is like a dashed relationship set because for both we must consult a lower level abstraction to fully understand the details.

Consider the dashed arrow in Fig. 4.12(b) that connects the high-level state *not packaging* to transition 9. A packager in the high-level *not packaging* state must be in one of the six low-level states or in transition between the states included within the cloud in Fig. 4.12(a). If a packager is in the *idle* state or the *clean up packaging room* state and the condition of transition 9 is true, the packager traverses the dashed arrow into the transition. However, if the packager is in any of the other four states (*on break, unable to get order, request boxes* or *request invoice forms*) and the condition of transition 9 is true, nothing happens. The packager cannot traverse the dashed arrow into the transition because in the lower level state net there is no arrow from any of these four states to transition *9.* If a packager is in one of the low-level transitions, the packager completes the low-level transition and enters one of the states. The action of the packager with respect to the condition of transition *9* depends on the lower level state entered. If the entered state is *idle* or *clean up packaging room,* the packager traverses the dashed arrow into transition *9.* Otherwise, nothing happens.

Similarly, the other two dashed arrows mean that a packager in the *not packaging* state may not always traverse them when their corresponding triggers hold. A packager traverses them only if the packager is in a state, or going into a state that, in the lower level state net, allows the transition.

Solid arrows incident on high-level states behave as solid arrows incident on atomic states. The arrows coming into the high-level state in Fig. 4.12(b) all declare that a packager enters the high-level state by entering at least one of the lower level states included within the high-level state. Similar statements can be made for the incoming arrows in Figs. 4.10(b) and 4.11(b). We never have dashed in-arrows.

Notice that in our previous examples, in Figs. 4.10(b) and 4.11(b), the out-arrows are solid. For Fig. 4.10(b) the out-arrow is solid because a packager in the high-level state must be in both of the subsumed low-level states, and thus when the condition of transition *10* is true, the packager always traverses the arrow. In Fig. 4.11(b) the packager may be in either state. However, the condition *package addressed* in transition *5* cannot be true unless the packager is in the lower level *address package* state. Thus when the condition of transition *5* holds, the packager traverses the arrow.

In general, an out-arrow from a high-level state is solid if an object in the high-level state always traverses the arrow when the trigger at the head of the arrow holds. Figure 4.13 gives an additional example. It shows the operation of a lamp with a three-way bulb. When we subsume the path from the *dim* state to the *medium* state to the *bright* state in the cloud in Fig. 4.13(a) as the high-level *on* state in Fig. 4.13(b), the out-arrow is dashed. The arrow is dashed because the lamp goes into the *off* state only when the lamp is in the *bright* state and the switch is activated. If the lamp is in the *dim* or *medium* states, the lamp does not turn off when the switch is activated.

4.5.2 Valid State Views

Whenever we create a high-level state, we wish to ensure that the diagram, after subsumption, is valid. As with the construction of high-level object classes and high-level relationship sets, we otherwise wish to allow the greatest possible freedom.

We have seen in the example of Fig. 4.10 that a high-level state need not include any transitions. When we include a transition, however, we must also include all its enabling and subsequent states. If we do not hold to this requirement, we could create high-level state nets with adjacent states and no intervening transition, which is not permissible.

We have not discussed subsumption of states under a dominant state. As for high-level relationship sets, we do not forbid dominant state construction, but we expect that it will be rare. Furthermore, dominant state construction does not impose any special requirements about how a dominant high-level state relates to its low-level states and transitions. Indeed, subsumption under a dominant state is the same as subsumption under an independent state name except that we reuse the name of some lower level subsumed state.

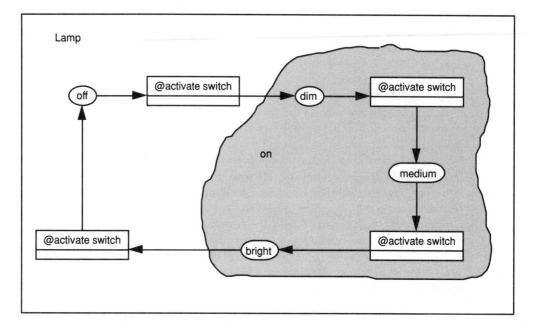

(a)

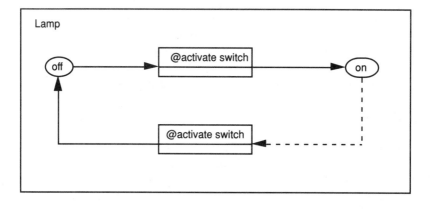

(b)

Figure 4.13   Limited exit from a high-level state.

4.6 High-Level Transition Views

A *high-level transition* groups transitions, states, constraints, and notes into a single transition. In Fig. 4.9 for example, we may not have much interest in a packager's leisure activities and may wish, therefore, to think of the time away from work as simply a transition away from and back to work. As shown by the cloud in Fig. 4.14(a), we can do

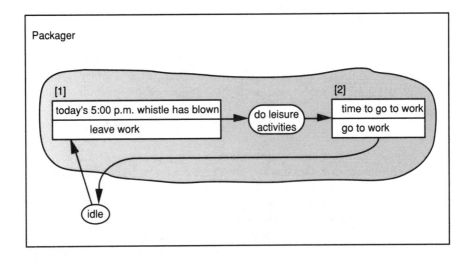

(a)

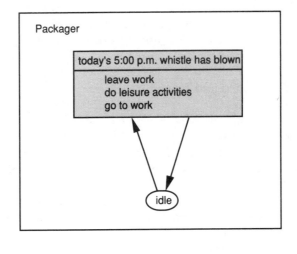

(b)

Figure 4.14   Construction of a high-level transition.

this by grouping transitions *1* and *2* and the *do leisure activities* state into a single high-level transition to and from work.

Figure 4.14(b) shows the result of imploding the high-level away-from-work transition. The transition is shaded to indicate that it is high-level. It is an independent high-level transition since it differs from any of the transitions inside the cloud. Indeed, most high-level transitions are independent. The transition has no transition number. We may add a number if we wish, but since they are not required we do not add one in this example. The arrows are all solid since they follow the defined conventions for arrow traversal. Indeed, high-level transitions have no effect on whether arrows are solid or dashed.

4.6.1 Triggers and Actions in Transition Views

An analyst who creates a high-level transition has the responsibility to provide its contents. For some simple constructions there are reasonable guidelines for determining the trigger and for deriving the actions. For a linear path of alternating states and transitions as in Fig. 4.14, the trigger can be the trigger of the entry transition, which for our example is

today's 5:00 p.m. whistle has blown.

The actions for a linear construction can be a sequence of the actions along the path from entry to exit. Actions along the path consist of actions in transitions and names of states. Thus, we can reasonably derive the actions of the high-level transition in Fig. 4.14(b) as being *leave work* from transition *1*, *do leisure activities* from the *do leisure activities* state, and *go to work* from transition *2*.

Figure 4.15 shows a parallel construction. Figure 4.15(a) shows that when a student is at home and it is time to go to school, the student goes to school either by driving or biking depending on whether it is raining or not raining. We can form a high-level transition from the two transitions as Fig. 4.15(b) shows. In a parallel construction, the trigger can be a disjunction of the triggers of all the included transitions. For our example, the trigger in the high-level transition fires if it is *time to go to school* and it is *raining* or if it is *time to go to school* and it is *not raining*, which simplifies to just if it is *time to go to school*. The actions in a parallel construction can be a disjunction of the actions of all the included transitions. For our example, the actions are *drive to school* or *bike to school*.

4.6.2 Exceptions and Transition Views

Sometimes we may wish to ignore the details of exception handling associated with a transition. If so, we can create a dominant high-level transition and subsume all the exception handling under it. The cloud in Fig. 4.16(a) encloses the dominant transition, transition *9*, and all the states and transitions required to handle any exception emanating from transition *9*. The result after subsumption is shown in Fig. 4.16(b).

Observe that the new high-level transition has the same trigger and set of actions as the original. Subsumption under a dominant transition retains the trigger and actions of the dominant transition.

There is one exception arrow emanating from the new high-level transition. The exception arrow has a derived exception condition. We derive exception conditions for an arrow leading to a state S as a disjunction of the exception conditions that can lead from a transition to S. Thus we have a disjunction of the three exception conditions from the original diagram on the single exception arrow in the new high-level view.

4.6.3 Valid Transition Views

As before, we wish to ensure that a view resulting from the construction and implosion of a high-level transition is valid, but we do not wish to otherwise constrain the choice of items that may constitute a high-level transition. The one requirement we have is that if a state is part of a high-level transition, then all its associated transitions must also be part of the high-level transition. Otherwise, we will have illegal adjacent transitions.

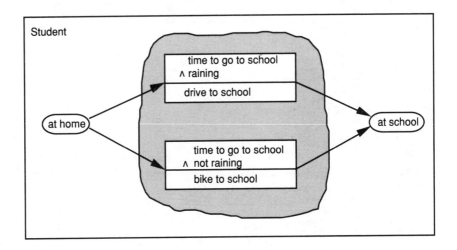

(a)

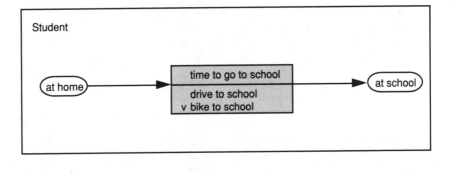

(b)

Figure 4.15   Parallel high-level transition construction.

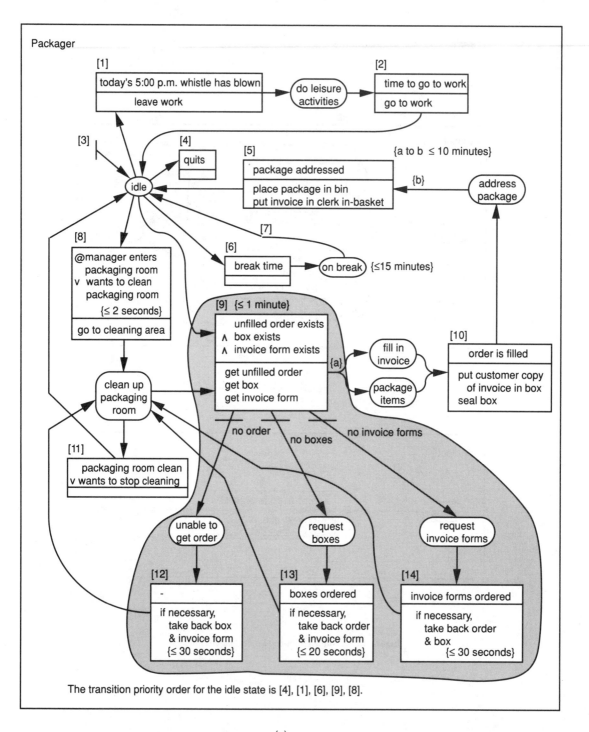

(a)

Figure 4.16 Construction of a dominant high-level transition.

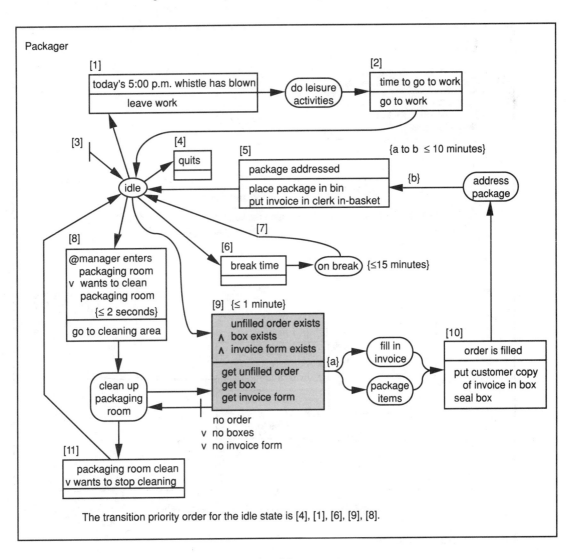

(b)

Figure 4.16 (Cont'd.)   Construction of a dominant high-level transition.

4.7 A High-Level State Net for a Packager

To conclude our section on high-level state-net views, we present a high-level view in Fig. 4.17 that nicely defines the behavior of a packager. It describes the main states of a packager, *not packaging* and *packaging*, and the transitions into and out of these main states.

We can create this high-level state net as follows. We may first create the *packaging* state as we did in Figs. 4.10 and 4.11. Next, we create the *not packaging* state as we did in Fig. 4.12. Finally, we create the high-level transition for activities as we did in Fig. 4.14.

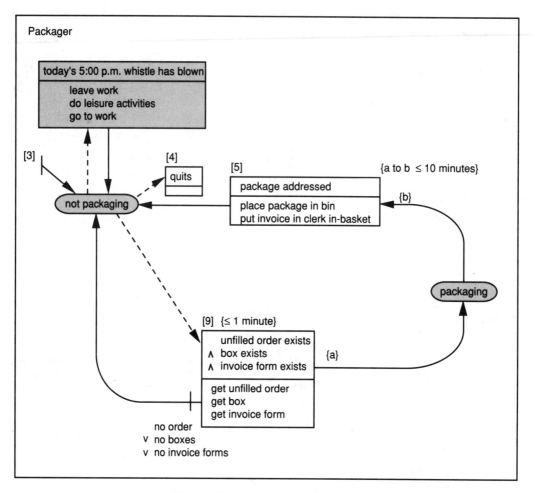

Figure 4.17　A high-level view of packager behavior.

4.8 Bibliographic Notes

Some of the abstraction mechanisms presented here are similar to abstraction mechanisms found elsewhere. Dominant high-level object classes, for example, are similar to those presented by Teorey [Teorey 1989] and high-level state views have some similarities with Harel's statecharts [Harel 1988]. OSA is unique, however, in that high-level views can be used to consistently structure all the modeling constructs in OSA.

Most techniques that are not object-oriented rely only on aggregation to organize their information. Aggregation is usually expressed as a hierarchical "subpart" relationship and is applied both for processes and data. Examples include PSL/PSA [Teichroew 1977], SADT, [Ross 1977], and SRD [Orr 1981]. Structured analysis uses leveling to structure data flow diagrams, but relies only on aggregation in data dictionaries to organize data [DeMarco 1979, Gane 1979, McMenamin 1984, Ward 1985, Yourdon 1989]. When semantic data models have been used with structured analysis, no abstraction mechanism has been introduced [McMenamin 1984, Ward 1985, Yourdon 1989].

Similarly, when state transition diagrams are added, often no abstraction mechanism is described [Ward 1985, Yourdon 1989].

Differing approaches for information organization are used in the various object-oriented analysis techniques. Coad and Yourdon use the classic forms of aggregation, classification, and generalization as found in extended ER models and object-oriented programming languages [Coad 1991]. They use the concepts of "problem domain and sub-domain," "subjects," and "layers." A "subject" is roughly similar to our notion of dominant high-level object class. The use of "layer" combinations to organize detail corresponds somewhat to our use of views. However, exactly how combinations of layers can be used is not brought to full definition within the text.

OMT provides the standard aggregation, classification, and generalization mechanisms of extended ER models [Rumbaugh 1991]. It does not, however, provide view mechanisms similar to those in OSA. For OMTs process-oriented interaction model, represented by data flow diagrams, leveling is used to structure the information. For OMTs behavior model, represented by statecharts, Harel's notion is used [Harel 1988]. Thus, OMT allows the partitioning of behavior models with respect to object classes and uses superstates and substates to organize complex behavior models.

4.9 Exercises

4.1 Figure 4.18 shows an ORM used for the KWE Bank. Use dominant high-level object classes to abstract and simplify the diagram as follows.
  a. Show low-level object classes and relationship sets and designate the high-level classes with a cloud similar to the cloud encapsulations shown in this chapter.
  b. Show the imploded high-level classes and any resulting relationship-set connections.

4.2 Suppose we were to create an independent high-level object class called Customer Information for the ORM in Fig. 4.18. Could the independent high-level object class contain all of the object classes and relationship sets in Fig. 4.18? Explain why or why not.

4.3 Figure 4.19 shows an ORM for part of the New Year's Club information. Create a high-level relationship-set specifying how a customer who is a member of the New Year's Club qualifies for a bonus interest rate.
  a. Show low-level object classes and relationship sets and designate the high-level relationship set with a cloud.
  b. Show the imploded high-level relationship set.

4.4 Figure 4.20 shows a partial state net for an ATM inquiry transaction. Simplify the behavior by creating three high-level states, one for setting up the transaction, one for doing the inquiry, and one for printing a receipt.
  a. Show the low-level state-net components of each high-level state in a cloud.
  b. Implode each high-level state and show the resulting state-net view.

4.5 As in Exercise 4.4 abstract and simplify the behavior depicted by the state net in Fig. 4.20. This time, however, create two high-level transitions, one for printing and one for processing the inquiry.
  a. Show the low-level state-net components of each high-level transition in a cloud.
  b. Show the two high-level transitions imploded and connected to the appropriate states from the original state net.

4.6 Using the partial state net in Fig. 4.21, create a dominant high-level transition that includes the transition and the invalid-PIN and stolen-PIN exceptions.

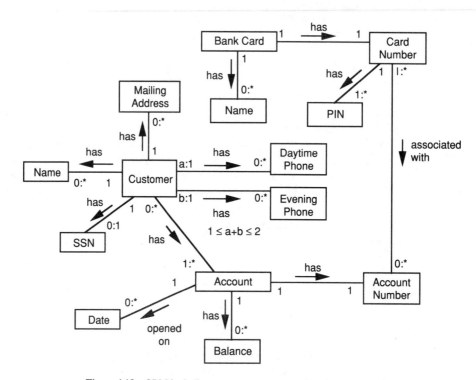

Figure 4.18   ORM including account, customer, and bank card objects.

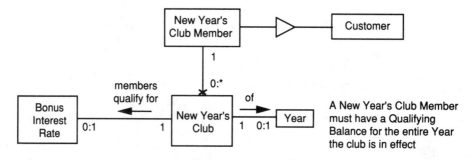

Figure 4.19   New Year's Club information.

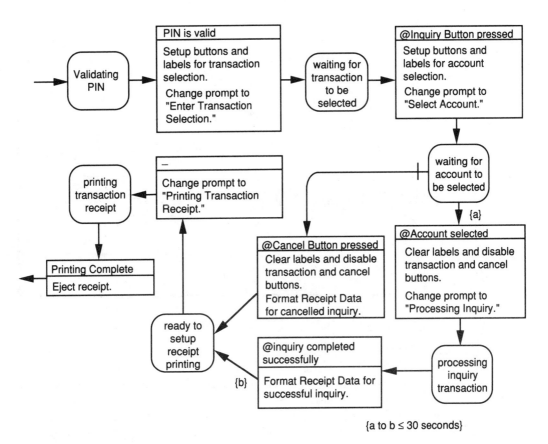

Figure 4.20   Partial state net for an ATM inquiry transaction.

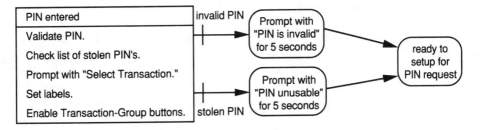

Figure 4.21   Partial state net showing invalid and stolen PIN exceptions.

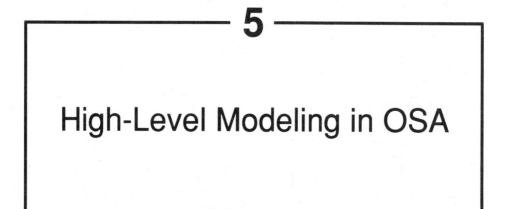

5

High-Level Modeling in OSA

In Chap. 4, we were interested in simplifying diagrams by creating high-level views of object classes, relationship sets, states, and transitions. In this chapter, we wish to consider high-level components as basic building blocks in OSA models. We wish, for example, to connect relationship sets directly to high-level object classes and to connect transitions directly to high-level states. To do so, we will need to establish the meaning of high-level components in terms of their low-level components.

In addition to addressing this main objective of semantically defining high-level components, we will also explore top-down model development. An analyst working to create an OSA model may initially create high-level components and then later provide additional detail by adding lower level components for an already created high-level component. In the packager state net of the Green-Grow Seed Company, for example, we may wish to provide a lower level state net for the *clean up packaging room* state that gives the details of what it means to clean up the packaging room.

The ability to group low-level components into high-level components to create OSA models provides an analyst with a bottom-up modeling perspective. The ability to treat existing components as high-level components and refine them provides an analyst with a top-down modeling perspective. Both top-down and bottom-up modeling are important, and each is appropriate for different tasks and even for different aspects of the same task.

5.1 High-Level Object Classes

High-level object classes are of two types, dominant and independent. Both types represent a set of objects as do atomic object classes. Dominant and independent object classes differ, however, in how they relate to the objects and relationships they represent. The set of objects represented by a high-level dominant object class is identical to the set of objects in the dominant object class. The set of objects represented by a high-level independent object class has a separate identity from the objects and relationships within the class. Although the independent objects have their own identity, they do have a definite correspondence to the underlying objects and relationships of the object classes and relationship sets included within the high-level object class. We explain this relationship in the discussion in Sec. 5.1.2 after completing our discussion of dominant high-level object classes in Sec. 5.1.1.

5.1.1 Dominant High-Level Object Classes

Figure 5.1 shows the construction of the dominant high-level object class *Order* from part of the ORM for the Green-Grow Seed Company. The set of objects in the low-level object class *Order* in Fig. 5.1(a) is identical to the set of objects in the high-level object class *Order* in Fig. 5.1(b).

We may attach a relationship set to a high-level object class. For example, we may begin with Fig. 5.1(b) and, as Fig. 5.2 shows, add the relationship set "*Customer registered Complaint about Order on Complaint Date.*" For a dominant high-level object class, attached relationship sets could, just as well, be attached to the low-level dominant object class. Because the objects within the low-level and high-level dominant object classes are identical, the meaning of the relationship set connected to the high-level dominant class is the same as if it were connected to the low-level dominant class. We do, however, keep track of whether it is the high-level object class or the low-level object class that is attached to the relationship set.

5.1.2 Independent High-Level Object Classes

Objects in independent high-level object classes intrinsically represent themselves as do all objects in every object class. Consider, for example, the independent high-level object class *Discount Information* in Fig. 5.3, which is constructed from part of the ORM for the Green-Grow Seed Company. To keep the example small, we assume that there are two preferred customer groups, one named *Blue Ribbon* with customers *Customer$_1$* and *Customer$_2$* and a *10* percent discount rate and the other named *Yellow Ribbon* with customer *Customer$_3$* and a *5* percent discount rate. We represent the discount information within the cloud by the table in Fig. 5.4(a). Each row in the table represents a conceptual discount-information object.

Letting *Discount Information$_1$* and *Discount Information$_2$* be the intrinsic identifiers for these conceptual discount-information objects, we can represent the relationship between the high-level discount-information objects and the low-level information comprising discount-information objects as the table in Fig. 5.4(b) shows. *Discount-Information$_1$* is related with the *Blue Ribbon* group, the customer set {*Customer$_1$*, *Customer$_2$*}, the discount rate *10%*, and the binary "*has*" relationships (*Blue*

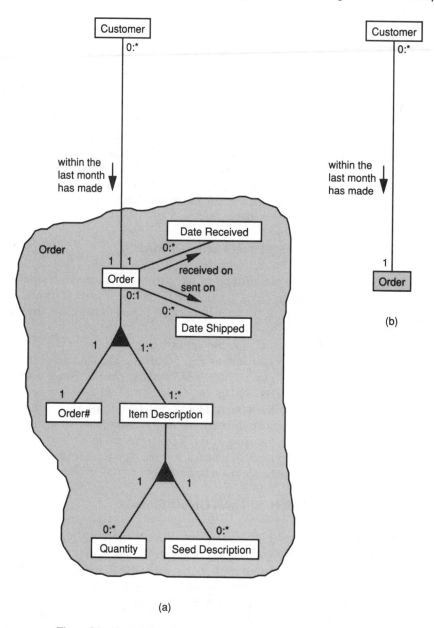

Figure 5.1   Construction of a high-level dominant object class for an order.

Ribbon, {*Customer₁, Customer₂*}) and ({*Customer₁, Customer₂*}, *10%*). Similarly, we
know the related objects and relationships for *Discount Information₂*.

The table in Fig. 5.4(b) also explicitly shows what we mean when we say that
independent high-level objects have an identity of their own, which is related to, but
separate from the information they represent. If we disregard all information in the table
except the object identifiers in the *Discount Information* column, we still are able to
identify the discount information objects in the high-level object class. Thus, each

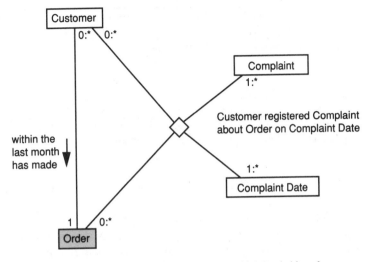

Figure 5.2   Addition of a relationship set to a high-level object class.

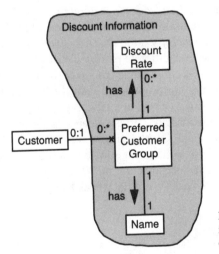

Figure 5.3   Construction of an independent high-level object class for discount information.

independent high-level object has an identity of its own. Also, we can see that each independent high-level object is related to low-level information as represented by a row in the table in Fig. 5.4(b).

An ORM provides a place to record the relationships between an independent high-level object class and its low-level components. This is done by specifying relationship sets that connect the high-level object class with one or more of its internal object classes. Figure 5.5, for example, shows the relationship set "*Discount Information goes with Preferred Customer Group*" connecting the high-level object class *Discount Information* to the internal low-level object class *Preferred Customer Group*. Notationally, when high-level object classes are not imploded and have internal connecting relationship sets, we connect the relationship-set lines to the edge of the cloud as Fig. 5.5 shows.

Name	Preferred Customer Group	Discount Rate
Blue Ribbon	{Customer$_1$, Customer$_2$ }	10%
Yellow Ribbon	{Customer$_3$ }	5%

(a)

Discount Information	Name	Preferred Customer Group	Discount Rate
Discount Information$_1$	Blue Ribbon	{Customer$_1$, Customer$_2$ }	10%
Discount Information$_2$	Yellow Ribbon	{Customer$_3$}	5%

(b)

Figure 5.4   High-level discount information objects.

With the *"Discount Information goes with Preferred Customer Group"* relationship set, we have enough information to define the relationship instances we saw in the table in Fig. 5.4(b). The participation constraints on *"Discount Information goes with Preferred Customer Group"* guarantee that a discount-information object connects to one and only one preferred customer group, which in turn has one and only one name and discount rate. We also could have connected *Discount Information* to *Name,* because the participation constraints guarantee the same meaning. However, we could not have specified the same meaning by connecting *Discount Information* to *Discount Rate* since a particular discount rate can be associated with several different preferred customer groups.

In the discount-information example we have been discussing, the high-level objects have a straightforward correspondence with the low-level objects and relationships. Each object corresponds to a row in a table representing the objects and relationships. In more complex situations, the correspondence is not so straightforward.

Consider, for example, the ORM in Fig. 5.6(a) in which *Car* is an independent high-level object class whose constituent low-level object classes are *Body, Motor, Wheel,* and *Spare. Spare* is a specialization of *Wheel.* Each car has at most one body and motor and up to five wheels, the fifth wheel being a spare. Many cars without

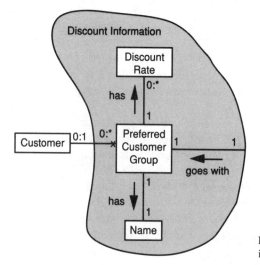

Figure 5.5   Internal high-level discount information connections.

bodies or motors and with less than five wheels can be found in auto junk yards, which is our sample application here.

The table in Fig. 5.6(b) shows a possible set of relationships among some cars, bodies, motors, wheels, and spares. Car_1 is intact since there is a body, a motor, four wheels, and a spare. Car_2 is just missing its spare. Car_3 has only its body and spare, and Car_4 has only its body and motor. Car_5 has its motor and a couple of wheels. Some people might not consider Car_5 to be a car, but our junk dealer does since these parts are what remain of Car_5. Our junk dealer would even consider something to be a car if it had no body, no motor, and no wheels; indeed this is the type of car our junk dealer likes best because all parts have been sold.

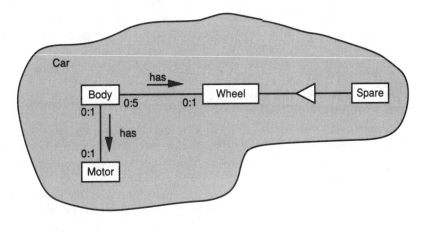

(a)

Car	Body	Motor	Wheel	Spare
Car_1	$Body_1$	$Motor_1$	$Wheel_1$ $Wheel_2$ $Wheel_3$ $Wheel_4$ $Wheel_5$	$Wheel_5$
Car_2	$Body_2$	$Motor_2$	$Wheel_6$ $Wheel_7$ $Wheel_8$ $Wheel_9$	
Car_3	$Body_3$		$Wheel_{10}$	$Wheel_{10}$
Car_4	$Body_4$	$Motor_3$		
Car_5		$Motor_4$	$Wheel_{11}$ $Wheel_{12}$	

(b)

Figure 5.6   High-level car objects.

In this example, the correspondence between a high-level object and its low-level objects does not form a single row in a table. However, we do have a partition among the objects. That is, each body, motor, wheel, and spare wheel belongs to one and only one car. The group of objects belonging to Car_2, for example, includes $Body_2$, $Motor_2$, $Wheel_6$, $Wheel_7$, $Wheel_8$, and $Wheel_9$. None of these objects belongs to another car. We can specify these restrictions by adding appropriate high-level to low-level relationship sets as the ORM diagram in Fig. 5.7 shows. The three additional relationship sets, "*Car has Body*," "*Car has Motor*," and "*Car has Wheel*" allow a car to have at most one body and motor and at most five wheels, and allow bodies, motors, and wheels to each belong to one and only one car.

Partitions among the low-level objects matter for some examples, but not for others. In our discount-information example, we do not have a partition among low-level objects because we allow several preferred-customer groups to have the same discount rate, but in our car example we maintain a strict partition among low-level objects. When there is a partition, we need only establish enough connections between the high-level object class and its contents so that we can follow internal relationships to find other objects and relationships associated with a high-level object. In our car example, we needed three connections to be sure that all necessary links to establish the desired meaning would be covered. Often, as in our discount-information example, we only need one link because the constraints guarantee that the internal components always connect throughout all internal object classes and relationship sets of a high-level object class.

When there is no partition, we cannot always rely on the underlying relationships alone to disambiguate the association of internal objects and relationships with objects in the high-level object class. Whether partitioned or not, we still follow internal relationships to find other objects and relationships associated with a high-level object. Often, as in the discount-information example, it does not matter that some objects have multiple associations with high-level objects. Sometimes it does matter, however, as we show in our next example. When multiple associations matter, the analyst must be more specific about which internal objects and relationships belong to a high-level object.

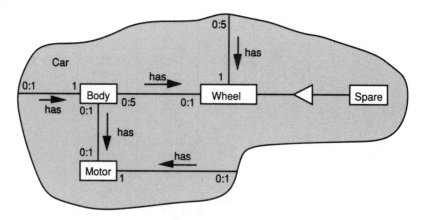

Figure 5.7   Internal high-level car connections.

Consider the ORM in Fig. 5.8(a) in which *Team* is an independent high-level object class with constituent low-level object classes *Coach, Player,* and *Position.* There are also other non-constituent low-level object classes *Team Name* and *Person. Team Name* connects to the high-level object class *Team* by the relationship set "*Team Name names Team.*" The tables in Fig. 5.8(b) give possible relationships among teams, players, and positions, which, for the example, we suppose are the true relationships. We assume here that the teams are softball teams from several different leagues. Perhaps *Team₁* is a church-league team and *Team₂* is a city-league team. We are thus able to have softball enthusiast *Person₄* play short stop for both teams, and enthusiast *Person₃* play pitcher and right field for *Team₁* and left field for *Team₂*.

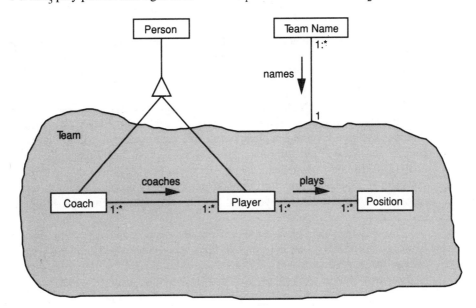

(a)

Team	Coach	Player
Team₁	Person₁	Person₃
		Person₄
		Person₅
	Person₂	Person₃
	. . .	
Team₂	Person₁₀	Person₃
		Person₄
		Person₁₁
	. . .	

Team	Player	Position
Team₁	Person₃	Pitcher
		Right Field
	Person₄	Short Stop
	Person₅	Catcher
	. . .	
Team₂	Person₃	Left Field
	Person₄	Short Stop
	Person₁₁	First Base
	. . .	

(b)

Figure 5.8   High-level team objects.

Unfortunately, the ORM in Fig. 5.8(a) cannot faithfully represent the information in the tables. No connection or combination of connections to internal object classes is enough to disambiguate the association of all internal relationships with a team object. In particular, we cannot always determine which positions a player plays, for a particular team. *Person₃* illustrates the problem. We cannot determine from the ORM model in Fig. 5.8(a) which position(s) *Person₃* plays for which team. If there were a connection between the high-level *Team* object class and the internal *Player* class the model could properly record that *Person₃* is both on *Team₁* and on *Team₂*, but we still would not know which "*Player plays Position*" relationship belongs with which team. Thus, for example, the model could incorrectly tell us that *Person₃* plays *Left Field* on *Team₁*, or it may incorrectly tell us that *Person₃* plays three different positions for *Team₁*—*Pitcher, Right Field,* and *Left Field.*

To solve this problem, we need to add more organizational information to the ORM. We need to be able to link internal relationship sets to high-level object classes. We can do this by making relational object classes for relationship sets that, in turn, link to the high-level object class through a relationship set. We then link any other low-level object classes to the high-level class as necessary.

In Fig. 5.9 we have added the relational object class *Player Position Assignment* linked to the high-level object class *Team.* We are now able to record that *Player₃* plays *Pitcher* and *Right Field* for *Team₁* and plays *Left Field* for *Team₂*. The participation constraints for the relationship set "*Player Position Assignment is for Team*" allow several player-position pairs for the same team so that a player can play several positions for the same team as does *Player₃* and can play the same position for different teams as does *Player₄*.

As this team example and the previous car example indicate, a correspondence can be established between an independent high-level object and many and possibly all the

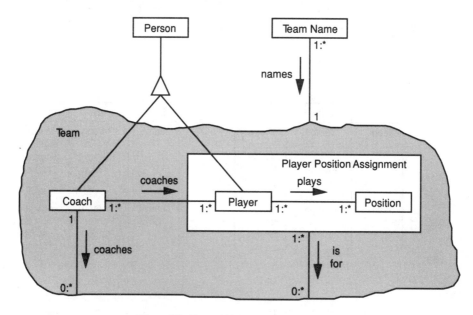

Figure 5.9   Internal high-level team connections.

objects and relationships it includes. We establish the correspondence by adding relationship sets that connect a high-level object class with included low-level object classes. Sometimes these low-level object classes may need to be relational object classes as in the latter example. The idea is to provide a way to specify, either directly or indirectly, which low-level objects and relationships belong to which high-level objects.

5.1.3 Similarities and Differences Among ORM Aggregation Mechanisms

A question naturally arises about the differences and similarities among aggregation relationship sets, relational object classes, and independent high-level object classes. Each of these abstraction mechanisms is similar because each groups constituent parts into a whole. However, there are also some differences.

A relational object class involves a single relationship set. We use it only when we wish to refer to a relationship in a relationship set as an object. For example, we may have *Couple* as the object class for the relationship set "*Husband is married to Wife*" or the *Address* object class for the relationship set "*Street Number, Street, City, State and Zip designate mailing Location.*"

An aggregation relationship set designates a parts assembly. We often use several aggregation relationship sets to form a hierarchy for a multilevel assembly. The top-level object class in the assembly denotes the assembled objects. Although relationship sets connecting subpart object classes in an aggregation relationship set are not forbidden, they are uncommon since the focus is on the assembly rather than on other possible interrelationships.

An independent high-level object class usually designates a complex submodel for a larger system model. Although we can use a high-level object class to model a relational object class or an aggregation relationship set, we prefer not to do so because these other specific aggregation mechanisms can more clearly capture the intent of the analyst.

5.1.4 Top-Down Perspective

We have been discussing high-level objects and their interrelationships with low-level objects and relationships. In developing high-level object classes, we may take a bottom-up approach in which we start with low-level object classes and relationship sets and then construct a high-level object class. We may also take a top-down approach in which we start with a high-level object class and then expand it into low-level object classes and relationships. In Chap. 4 we described bottom-up composition. Here we describe and illustrate a top-down perspective.

As might be expected, in our top-down perspective we reverse the order in which we create high-level and low-level components. There are two types of top-down creation. One corresponds to dominant high-level object classes, and the other corresponds to independent high-level object classes.

We can use Fig. 5.1 to illustrate top-down dominant high-level object class construction. We begin with the ORM in Fig. 5.1(b) where *Order* is a high-level object class. We then add relationship sets directly connected to *Order* and expand the part-of

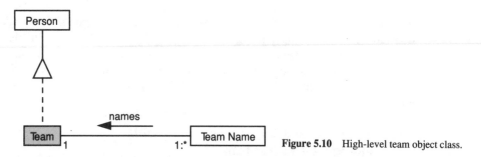

Figure 5.10 High-level team object class.

hierarchy rooted at *Order* as Fig. 5.1(a) shows. Since the objects in a dominant high-level object class are identical to the objects in the dominant object class in its low-level expansion, we expand high-level dominant object classes by adding additional detail as directly connected relationship sets and as subordinate aggregation and specialization hierarchies.

Figure 5.10 illustrates top-down independent high-level object class construction. We begin as Fig. 5.10 shows with the high-level *Team* object class. We consider *Team* to represent a complex subsystem with many interrelated object classes. Therefore, we create an ORM that represents *Team* and connect it as necessary to already-defined object classes such as *Person* in Fig. 5.10. The previous Fig. 5.9 shows a possible expansion of the high-level *Team* object class.

Our particular expansion of *Team* in Fig. 5.9(a) is simple enough that we may choose instead to treat *Team* as a dominant object class with related object classes for coaches, players, and positions. Even this choice begins to cause some problems, however, because it is unclear whether the relationship sets *coaches* and *plays* are part of *Team*. It becomes especially difficult and not advisable if we assume a more complex structure for *Team* that also includes equipment, inventory information for equipment including ownership and condition of equipment, equipment managers, and so on. Therefore, as is the case with any top-down approach, careful and knowledgeable choices must be made at each level of decomposition.

Independent high-level object classes often represent large, complex subsystems. These are particularly suitable for top-down development because we are not likely to understand all the complex structure of a subsystem when we begin the analysis.

5.2 High-Level Relationship Sets

High-level relationship sets represent sets of relationships as do atomic relationship sets. As for atomic relationship instances, we require that a high-level relationship instance be an *n*-tuple where *n* is the number of connections to associated object classes and where each object in the *n*-tuple is drawn from the object class to which it is connected.

A high-level relationship set represents a particular summarization of its low-level object classes and relationship sets. The purpose of the summarization may be to simplify the system model or to construct a specific view from existing low-level model components.

In general, an analyst gives the summarization for a high-level relationship set as a derivation in terms of its low-level object classes and relationship sets. Analysts may express these derivations as general constraints. Any derivation is acceptable, but two types of derivations are particularly common for high-level relationship sets. One is construction of the high-level relationship set by set operators, and the other is construction by composition. We discuss and illustrate these two common types of derivations in the following sections.

5.2.1 Construction by Set Operators

When several different relationship sets each have the same connections to a group of object classes, we may derive high-level relationship sets from low-level relationship sets by set operators. Figure 5.11(a) shows an example of three relationship sets each connecting to the same two object classes. A person may contribute financially to a symphony orchestra, may have season tickets for a symphony orchestra, and may do volunteer work for a symphony orchestra.

The cloud in Fig. 5.11(a) defines a high-level relationship set "*Person is patron of Symphony Orchestra.*" The general constraint in Fig. 5.11(a) specifies that the high-

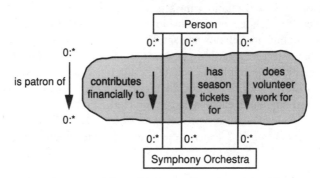

is patron of = contributes financially to U does volunteer work for U has season tickets for

(a)

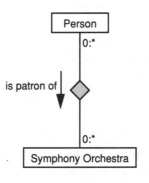

(b)

Figure 5.11   Specification of a high-level relationship set.

level relationship set is a union of the three low-level relationship sets. Thus, in the high-level abstraction in Fig. 5.11(b) a person is related to a symphony orchestra if the person is related to the symphony orchestra in any one, two, or three of the low-level relationship sets.

Other derivations are also possible. If, for example, we wish to have contributors to a symphony orchestra who neither do volunteer work nor have season tickets for the orchestra, we may give the constraint *is patron of = contributes financially to − (does volunteer work for ∪ has season tickets for).* As another example, we might have the intersection of all three low-level relationship sets and thus have only people who contribute to, do volunteer work for, and have season tickets for a particular symphony orchestra.

5.2.2 Construction by Composition

Construction by composition is assumed to be the default for high-level relationship sets in the absence of general constraints that specify the derivation. To introduce construction by composition, we use an example from the Green-Grow Seed Company. From the object class and relationship sets within the cloud in the ORM of Fig. 5.12(a), we construct the high-level relationship set *"Packager has same hourly rate as Clerk."* After imploding this relationship set, Fig. 5.12(b) shows the result. Figure 5.13(a) shows a possible object-relationship graph for an instance of the ORM in Fig. 5.12(a). In the object-relationship graph, we denote the three packagers as P_1, P_2, and P_3, and the five clerks as $C_1, \ldots, C_5$. We show the hourly rate each packager and clerk has by connecting a dollar amount to each packager and clerk. Figure 5.13(b) shows the instance graph for the high-level relationship set *"Packager has same hourly rate as Clerk."* The instance graph in Fig. 5.13(b) is the standard composition of the *"Packager has Hourly*

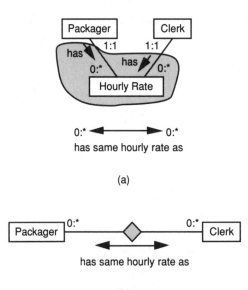

(a)

(b)

Figure 5.12   A high-level relationship formed by composition.

Rate" relationship set and the converse of the "*Clerk has Hourly Rate*" relationship set in Fig. 5.13(a). (We say converse here because the direction of the "*has*" relationship set is from *Clerk* to *Hourly Rate,* whereas we need the relationship set ordered from *Hourly Rate* to *Clerk.*) The relationship (P_1, C_1), for example, is in the result in Fig. 5.13(b) because it is the composition of $(P_2, \$9.35)$ in "*Packager has Hourly Rate*" and $(\$9.35, C_1)$ in the converse of "*Clerk has Hourly Rate.*" Similarly, (P_2, C_2) is the composition of $(P_2, \$9.35)$ and $(\$9.35, C_2)$, and (P_3, C_4) is the composition of $(P_3, \$13.50)$ and $(\$13.50, C_4)$.

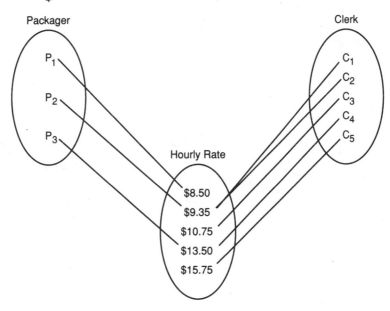

(a)

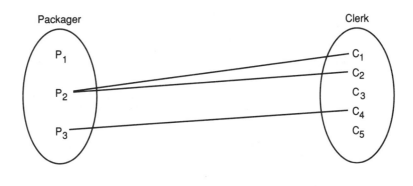

(b)

Figure 5.13   Object-relationship graphs for a high-level composition relationship set.

The composition for the high-level relationship set in Fig. 5.13(b) is straightforward, but the composition can be more complex. Consider as our next example the high-level relationship set "*A Customer belonging to the group named Name gets Discount Rate*" in Fig. 5.14. Figure 5.14(a) shows the initial construction of the high-level relationship set, and Fig. 5.14(b) shows the imploded high-level relationship set.

Figure 5.15(a) shows an object-relationship graph for a possible instance. For the object and relationship instances in Fig. 5.15(a), Fig. 5.15(b) shows the relationship instances in the high-level relationship set. We generate these high-level relationship instances from the low-level relationships using a generalization of composition.

We may describe the generalization of composition we use by describing the paths in an object-relationship graph. In the object-relationship graph in Fig. 5.15(a) there is a

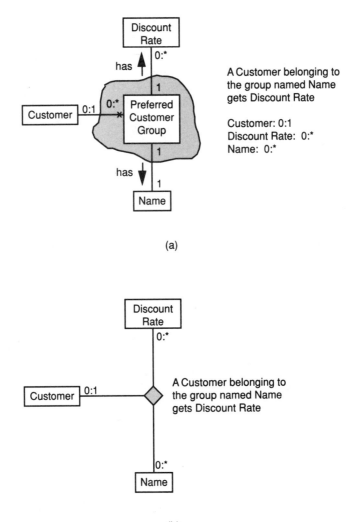

Figure 5.14   An *n*-ary high-level relationship set.

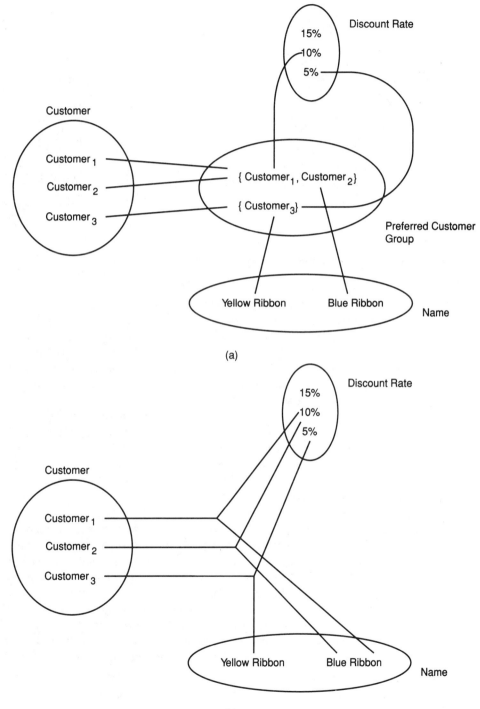

Figure 5.15 Object-relationship graphs for a high-level ternary relationship set.

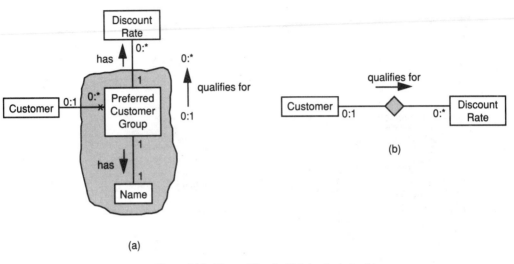

(a)

(b)

Figure 5.16   The qualifies-for high-level relationship set.

path through $\{Customer_1, Customer_2\}$ that links $Customer_1$, *10%*, and *Blue Ribbon*. Thus $(Customer_1, 10\%, Blue Ribbon)$ is a relationship in the high-level relationship set "*Customer belonging to the group named Name gets Discount Rate.*" To obtain the composition in general, we form a cross product of the sets of values in the n object classes of the high-level relationship set and keep only those n-tuples for which there is a connection throughout the object classes and relationship sets included within the high-level relationship set.*

As another example, consider the high-level relationship set "*Customer qualifies for Discount Rate*" in Fig. 5.16. For the objects and relationships in Fig. 5.15(a), Fig. 5.17 shows the result of the composition. We see in Fig. 5.17 that $Customer_1$ and $Customer_2$ qualify for a *10%* discount and that $Customer_3$ qualifies for a *5%* discount.

5.2.3 Other Constructions

Although construction by set operations and by composition are common, an analyst may specify other constructions. To illustrate one possibility, we may wish to have "*Customer qualifies for excellent Discount Rate*" in our last example instead of "*Customer qualifies for Discount Rate*" where "*Customer qualifies for excellent Discount Rate*" contains only those high-level relationship sets in which the discount rate is greater than 10 percent. We may restrict our composition with the general constraint "*Restrict qualifies-for to have only Discount Rates > 10%.*" For our sample data in Fig. 5.15(a), the set of relationship instances in "*Customer qualifies for excellent Discount Rate*" would be empty because no customer qualifies for a discount rate greater than 10 percent.

---

* The reader acquainted with relational database systems will recognize the composition we are discussing as a natural join followed by a projection.

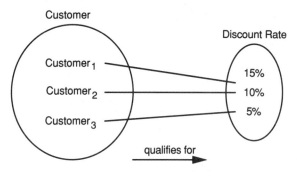

Figure 5.17   Object-relationship graph for the high-level qualifies-for relationship set.

5.2.4 Top-Down Perspective

As with high-level object classes, we may take a top-down perspective of relationship-set construction by considering the high-level relationship sets first. We may, for example, first have the high-level *qualifies for* relationship set shown in Fig. 5.16(b) and then later supply its details. The details consist of connecting object classes and relationship sets that explain why a customer qualifies for the discount rate. As Fig. 5.16(a) shows, a customer qualifies for a discount rate because of membership in a preferred customer group and because each customer in a preferred customer group qualifies for the discount rate of the group.

5.3 High-Level States

When the activity performed in a state is defined by another state net, the state is a high-level state. We may define high-level states bottom-up by creating high-level states from low-level states and transitions. We may also define high-level states top-down by providing a state net for an existing state.

5.3.1 Top-Down Development of State Details

We first consider top-down development by providing a state net for an existing state. Since we will make several references to the *Packager* state net, we again provide it as Fig. 5.18.

Figure 5.19 is a state net for the *clean up packaging room* state of Fig. 5.18. We choose to denote this by placing the substate net inside the *clean up packaging room* state as shown in Fig. 5.19. The state net describes the activities involved in cleaning a packaging room. A packager first assesses the situation and then either sweeps the floor, straightens the room, or takes out the trash. After completing any of these activities, the packager reassesses the situation, and continues cleaning the room as necessary.

There are several interesting features in this low-level state net for cleaning up the packaging room. We are interested, in particular, in how a packager enters and exits the low-level state net and how the low-level state net relates to the high-level state net that includes the state whose activity is described.

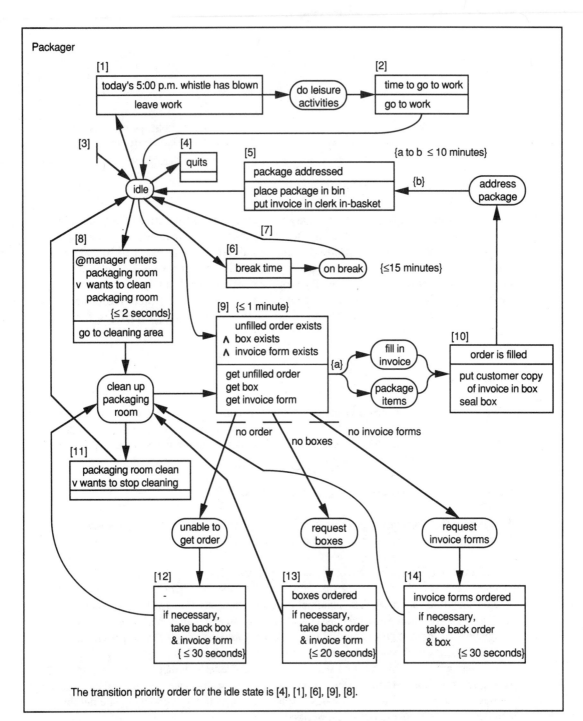

Figure 5.18 Packager state net.

5.3.1.1 Entry specification for low-level state nets.   When an object enters a high-level state, it always enters at least one low-level state in the state net that defines the activity of the high-level state. There are several ways we may denote the entry states. We show one way in Fig. 5.19 where we designate that a packager enters the *assess situation* state by an in-arrow with no attached transition. From Fig. 5.18, we know that a packager may enter the *clean up packaging room* state from five different transitions. Upon entering the high-level state from any of these transitions, the in-arrow in Fig. 5.19 designates that a packager first enters the low-level *assess situation* substate.

We choose to let in-arrows with no attached transitions designate initial states for state nets of high-level states. If an object is to enter several states concurrently, the initial in-arrow must branch off to several heads, one for each of the concurrently entered states. If there are several separate initial in-arrows in a state net, an object may arbitrarily choose any one as the initial in-arrow. As usual, constraints may be added so that the choice is not arbitrary.

5.3.1.2 High-level exit from low-level state nets.   We next consider how a packager exits a low-level state net. Observe that the state net for the *clean up pack-*

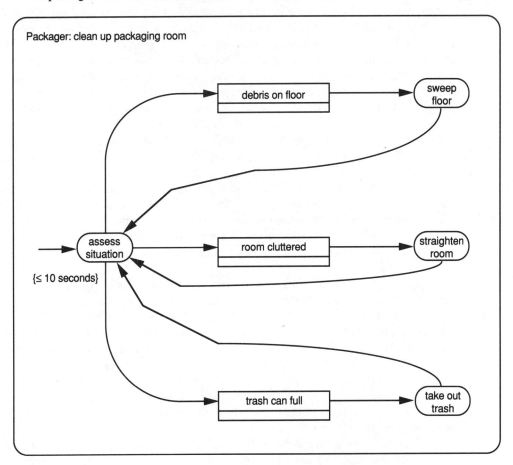

Figure 5.19   A state net for cleaning up the packaging room.

aging room state of a *Packager* in Fig. 5.19 allows a packager at most 10 seconds to assess the situation and begin to take action. However, if the floor is swept, the room is straightened, and the trash has been taken out, no transition trigger holds, and thus a packager would necessarily have to violate the real-time constraint on *assess situation*. Thus, because of the real-time requirement, the state net would not make sense if it stood alone.

Since the state net is a low-level state net for the high-level *clean up packaging room* state in Fig. 5.18, the trigger in transition *11*,

$$packaging\ room\ clean\ \lor\ wants\ to\ stop\ cleaning$$

and the trigger in transition *9*,

$$unfilled\ order\ exists\ \land\ box\ exists\ \land\ invoice\ form\ exists$$

may cause the packager to leave the *clean up packaging room* state, and therefore, leave all states in the lower level state net. In particular, when the packaging room is clean, the trigger in transition *11* of Fig. 5.18 holds, and thus a packager need not violate the real-time constraint by remaining in the *assess situation* state longer than 10 seconds.

If there is a solid out-arrow from a high-level state to an external transition and the trigger of the transition holds, an object leaves the high-level state. When an object leaves a high-level state, it leaves all low-level states within the high-level state including any low-level states that are themselves high-level states. If a low-level transition of a high-level state for an object is active, the low-level transition completes before the high-level state is exited. If the external transition fires and is connected to the high-level state by an exception in-arrow, conceptually all states and transitions are exited immediately. For exceptions, we do not wait for the low-level transitions to complete before exiting the high-level state.

When a packager is in the *sweep floor* state in Fig. 5.19 and the trigger of transition *9* in Fig. 5.18 holds, for example, the packager leaves the *clean up packaging room* state to fill an order and thus also leaves the *sweep floor* state. We can also see that because of transition *11* a packager may leave any of the four substates in the *clean up packaging room* state because the packager simply wants to stop cleaning.

5.3.1.3 Controlled entry into low-level state nets.

As we have defined the *clean up packaging room* state of the *Packager* state net, a packager always enters in the *assess situation* state. To illustrate some other possibilities, let us assume that when an idle packager "jumps up" to start cleaning the packaging room because the manager enters the room, that the packager always grabs a broom and starts sweeping. Hence, a packager entering the *clean up packaging room* from transition *8* in Fig. 5.18, should always directly enter the *sweep floor* substate. Let us further assume that if a packager is cleaning the packaging room, is interrupted to fill an order, and is unable to fill the order because of an exception condition, then the packager should resume cleaning the packaging room in the state exited.

To control entry from a particular transition, we provide a generalization of our initial state in-arrow. We generalize by allowing a transition-entry specification to be added to the tail of an in-arrow. A transition-entry specification is a list of transition identifiers designating the high-level transitions from which a substate may be entered. The list limits the possible transitions into the substate to come only from a transition on the list. An initial state with no transition-entry specification has no limitation and thus may come from any of the high-level transitions.

Figure 5.20 shows an example. The diagram in Fig. 5.20 is the state net of Fig. 5.19 altered to allow direct entry into the *sweep floor* state, but only from transition *8*, which is the transition triggered by the manager entering the packaging room. The state net is also altered to allow entry into the *assess situation* state, but only from transitions other than transition *8*. Thus, a packager entering from transition *8* enters the *sweep floor* state and from all other transitions enters the *assess situation* state.

For entries into high-level states that cannot be specified by initial state in-arrows with added transition-entry specifications, an analyst may give an entry specification by adding general constraints to the state-net diagram. The constraints that define the entry

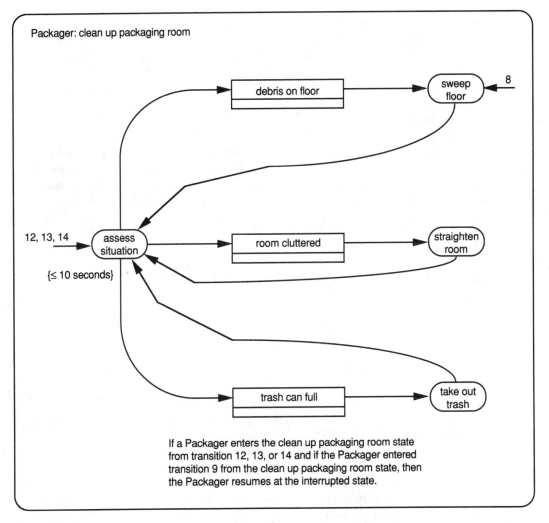

Figure 5.20 A state net with entry specifications.

specification may depend on any aspect of the current model instance including any historical information stored for the specification. An item of historical information that is typically of interest is the set of interrupted substates for an object, which is empty if the object has not yet entered the high-level state.

For our revised example, where we wish to have a packager re-enter an interrupted state after exiting transition *9* because of an exception condition, we add the following constraint given in Fig. 5.20.

> *If a Packager enters the clean up packaging room state from transition 12, 13, or 14 and if the Packager entered transition 9 from the clean up packaging room state, then the Packager resumes at the interrupted state.*

With this constraint and the transition-entry specifications added as in Fig. 5.20, we are able to specify the desired behavior.

For some situations we may need to keep track of several different interruptions. Figure 5.21 gives an example. In Fig. 5.21(a) we see that an *Order Taker* can be in a high-level state called *taking orders*. The low-level state net that describes the details of this state are in Fig. 5.21(b). While in any of the substates of *taking orders*, another call can come in. When a call does come in, the order taker puts the person making an order on hold, answers the in-coming call, and re-enters the *taking orders* state. At this point, the trigger in transition *4* is satisfied and hence the order taker puts the new caller on hold and reconnects the previous caller. Because the order taker can be interrupted many times before finishing a call, there may be several calls on hold when a person finishes placing an order. When an order is complete, transition *5* fires if anyone is on hold; otherwise transition *2* fires. The actions in transitions *4* and *5* ensure that the oldest caller has the highest priority. The added entry constraint in Fig. 5.21(b) allows the order taker to resume taking orders properly from someone who had been on hold.

If we wish to be specific about the historical information needed for re-entry into a low-level state net, we may model it in an ORM. In our last example, we would need to record the state of each caller on hold.

5.3.2 Bottom-Up Development of High-Level States

Having discussed top-down creation of a state net for an existing state, we now turn our attention to creating high-level states bottom-up from existing low-level states and transitions. We will be particularly interested in entry and exit specifications.

5.3.2.1 Exit specification for low-level state nets.   Consider the state-net creation illustrated in Fig. 5.22. Figure 5.22(a) shows the full state net with a cloud covering the high-level *on* state, which has been imploded in Fig. 5.22(b). Figure 5.22(c) shows the low-level state net for the high-level *on* state. Observe that in addition to the special notation that designates the *dim* substate as the initial state in Fig. 5.22(c), we also have special notation for the *bright* substate. We choose to let a single out-arrow emanating from a state and not connecting to a transition designate that a substate is a final state from which we expect to exit the state net.

In Chap. 4 we discussed the meaning of a dashed arrow. In Fig. 5.22(b), the arrow is dashed because the lamp does not necessarily leave the *on* state when the switch is activated. The lamp only leaves the *on* state when the switch is activated and the lamp is

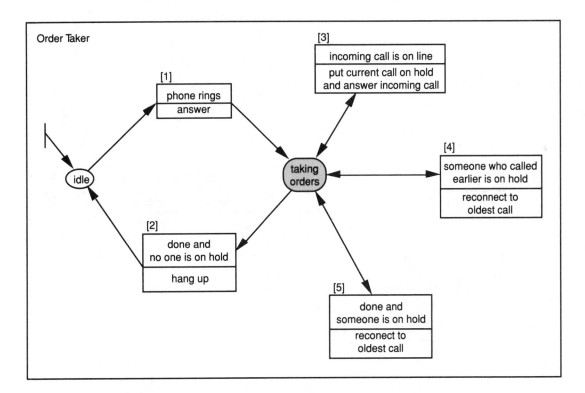

(a)

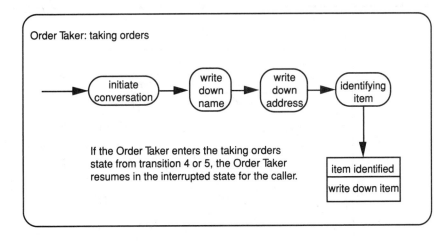

(b)

Figure 5.21   A low-level state net with multiple resume states.

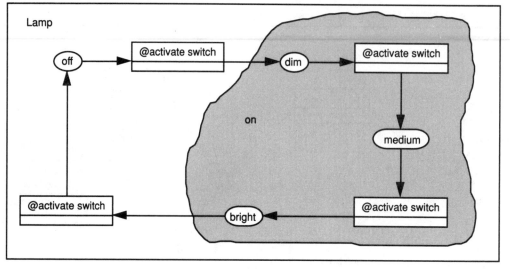

(a)

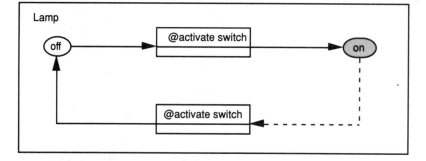

(b)

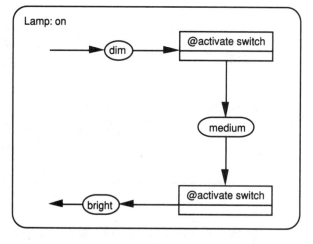

(c)

Figure 5.22   High-level state net with a final state.

in the *bright* substate as Fig. 5.22(a) shows. From the low-level state net point of view, there is a dashed arrow because there are states from which the lamp does not leave the high-level *on* state even though there is a transition connected to the high-level *on* state whose trigger may be satisfied.

In low-level state nets we designate final states from which exits along dashed arrows may occur by using solid out-arrows. Final state out-arrows have multiple tails when an object concurrently exits from several states. If it is possible for an object to be in a final state and a nonfinal state concurrently, then, in the absence of special constraints, it must also be possible for the object to be concurrently in both the high-level state and the state(s) to which the object exits from a final state.

5.3.2.2 Controlled exit from low-level state nets.   Sometimes we may wish to limit the transitions that may trigger an exit along a dashed arrow. Similar to transition-entry specifications, we may place transition-exit specifications on out-arrows to limit the transitions that may affect final states.

We show an example in Fig. 5.23. Figure 5.23(a) shows the bottom-up creation of the *Packager not packaging* state net in Fig. 5.23(b). Figure 5.23(b) shows that a packager in the low-level *clean up packaging room* state may only leave the high-level state through transition *9* of Fig. 5.23(a).

Similar to transition-entry specifications, when there is no transition-exit specification, an object may leave a final transition through every transition that follows the high-level state. For example, a packager in the *idle* state in Fig. 5.23(b) may leave through transition *1, 4,* or *9*, which are all the transitions that follow the high-level state as can be seen in Fig. 5.23(a).

5.3.2.3 Entry and exit specifications and boundary-crossing connections.   We may consider transition-entry and -exit specifications as specifications of boundary-crossing connections between a low-level state net and the high-level state net in which it is embedded. We see for example in Fig. 5.23(a) that there are four transitions, *2, 3, 5,* and *9*, from which there are boundary-crossing arrows into the high-level *not packaging* state. Since only transitions *2, 3,* and *5* connect to the *idle* state, its transition-entry specification limits entry from only transitions *2, 3,* and *5*. Similarly, the transition-entry specifications for the *unable to get order* state, the *request boxes* state, and the *request invoice forms* state, limit entry to only transition *9*.

Observe in Fig. 5.23(b) that exception arrows carry their conditions with them. This makes it possible to know which substate is entered when a particular exception occurs.

Transition-exit specifications similarly reflect boundary-crossing out-arrows. In Fig. 5.23(b) the out-arrow from the *clean up packaging room* state limits the transition to *9* because this is the only transition to which the arrow emanating from the *clean up packaging room* connects. The out-arrow from the *idle* state is unlimited because there is a connection to all transitions to which boundary-crossing out-arrows connect.

When we create state nets for high-level states bottom-up, transition-entry and -exit specifications are already known, and therefore, need not be specified. When we create state nets top-down, transition-entry and -exit specifications allow us to designate the connections. We may choose not to display transition-entry and -exit specifications when they are of no concern for our application.

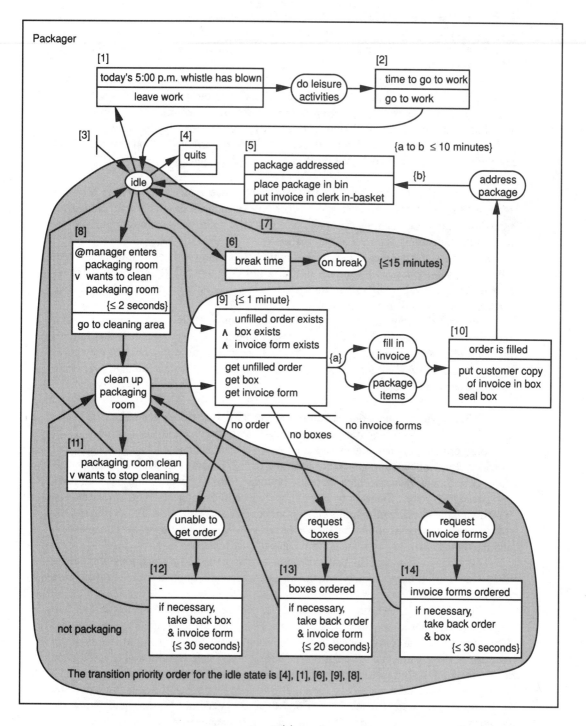

(a)

Figure 5.23 State net for the high-level not packaging state of packager.

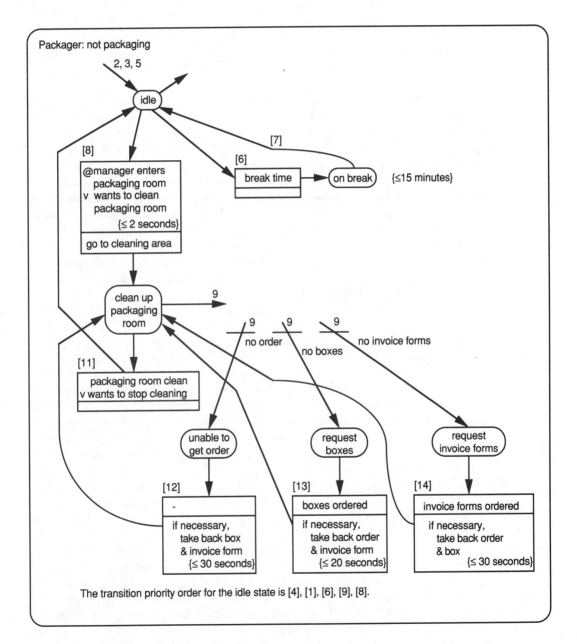

(b)

Figure 5.23 (Cont'd.)   State net for the high-level not packaging state of packager.

5.3.3 High-Level States and Aggregations

We have found that the high-level-state concept provides a convenient mechanism for viewing the behavior of aggregate objects. The behavior of an aggregate object is a conglomerate of the behaviors of each of its subparts and the global behavior that ties all the subparts together. It is often difficult to separate the behavior of the aggregate from its subparts. We can, however, find the states and transitions that correspond to a subpart and use a high-level state to represent them. These high-level states may be displayed as part of the aggregate state net or displayed separately, as the analyst deems best.

To illustrate these ideas we present an example of a simple stopwatch. Figure 5.24(a) shows the aggregate *Stopwatch* and its two subpart object classes *Timer* and *Display*. Figure 5.24(b) shows the full state net for the stopwatch. Notice that the *on* state in Fig. 5.24(b) is a high-level state and is displayed with its contents and in its context as part of the state net. The stopwatch has two buttons: a main button and an alternate button. The first time the main button is pushed, the time is set to zero and the stopwatch enters both the *time is running* state and the *displaying MM:SS:TT* state. When the main button is pushed again, timing stops, and when it is pushed once more, the time is set to zero and time begins running again. Independently, the alternate button may also be pushed. The alternate button causes the display to switch alternately between displaying minutes, seconds, and tenths of seconds and displaying hours, minutes, and seconds. If the stopwatch remains in the *time is stopped* state for more than five minutes, the stopwatch automatically turns off.

To understand the behavior of the subparts of the stopwatch, we first identify the states and transitions associated with each of the subparts. Although not necessary here, we may need to omit some initial or final transitions to make the set of states and transitions for a subpart correspond to a high-level state. We then form a high-level state for each subpart. Figure 5.25 shows the result for our example. Figure 5.25(a) is the high-level *timing* state that corresponds to the *Timer* subpart, and Fig. 5.25(b) is the high-level *displaying* state that corresponds to the *Display* subpart. Figure 5.25(c) shows the aggregate state net with these new high-level states imploded. Alternatively, we may impose these high-level states directly on the diagram in Fig. 5.24(b) without imploding them.

Since these high-level states correspond directly to object classes in an aggregation, we also have one more way we can display these state nets. Figure 5.26 shows that we can also display the subpart object class. In Fig. 5.26(a) we have merely added the object class *Timer,* but in Fig. 5.26(b) we have also omitted the high-level state symbol for the *displaying* state. In this context, the high-level state symbol is redundant and may be displayed or omitted as desired.

As one last way to view the behavior of an aggregation with respect to its subparts, we show a combined diagram in Fig. 5.27. Here, we recognize that high-level states for each of the subparts have been created, but instead of displaying the high-level states, we display the rectangular object classes to which the high-level states correspond. The diagram in Fig. 5.27 nicely shows which part of the behavior corresponds to the two subparts *Timer* and *Display* and which part of the behavior corresponds to the aggregate *Stopwatch*. It also nicely shows how the behavior of the subparts and the behavior of the aggregate tie together.

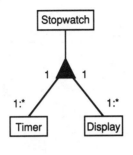

(a)

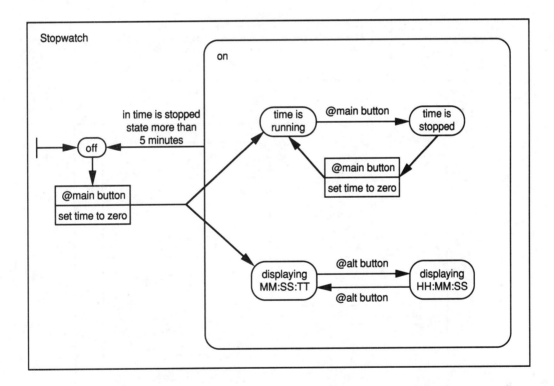

(b)

Figure 5.24   Stopwatch as an aggregate of timer and display.

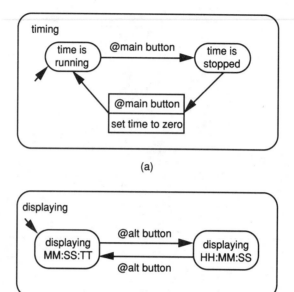

(a)

(b)

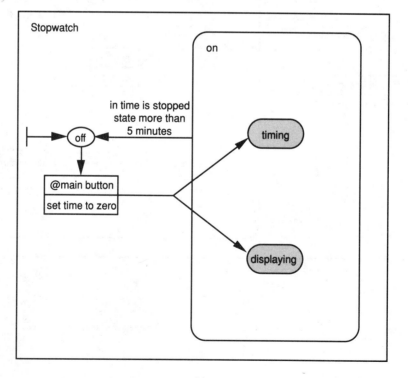

(c)

Figure 5.25   The timing and displaying high-level states.

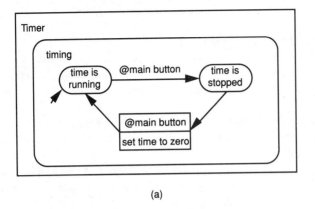

(a)

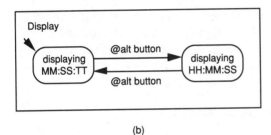

(b)

Figure 5.26   Subpart object classes for the timing and displaying high-level states.

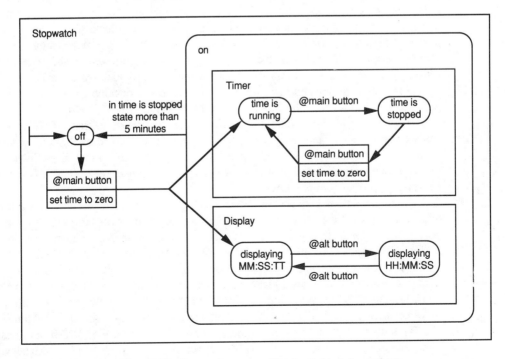

Figure 5.27   Stopwatch state net with subpart object classes shown.

5.4 High-Level Transitions

As for high-level states in state nets, we may create high-level transitions either top-down or bottom-up. For top-down creation, an analyst supplies a state net that describes the behavior of a high-level transition, and for bottom-up creation, an analyst creates a high-level transition by specifying which existing state-net components belong to the new transition.

As with low-level state nets for high-level states, one of our main concerns is the entry into and exit from high-level transitions. To address these concerns, we provide a representation for context information needed to understand low-level state nets for high-level transitions. The context information consists of entry and exit specifications. The entry and exit specifications are similar to the entry and exit specifications of low-level state nets for high-level states. We will see, however, that there are some additional problems to solve, due mainly to exceptions exiting from high-level transitions.

We begin our discussion on state nets for high-level transitions by discussing top-down transition creation, in which we provide a state net for a transition where none previously existed. We then discuss bottom-up transition creation. In our discussion of bottom-up creation, we also explain our conventions for specifying entry into and exit from a state net for high-level transitions.

5.4.1 Top-Down Development of Transitions

As an example of top-down transition creation, we provide a state net for transition *10* in the *Packager* state net, which we show in Fig. 5.28(a). The given action part of transition *10* states that when an order is filled, a packager puts the customer copy of the invoice in the box and then seals the box. The state net we create should define how a packager performs these actions. Since the action description is never required to be formal, we are unable to guarantee semantic equivalence between the statements in the action part of the high-level transition and the state net provided by an analyst. This is typical of top-down development, where, as we refine and develop a system, we usually provide more formal detail for less formal high-level statements.

Figure 5.28(b) shows a possible state net for the actions in transition *10*. Notice that we place the sub-state net in a transition rectangle as denoted by the additional horizontal line near the top of the rectangle. We see in Fig. 5.28(b) that the actions in the transition can be represented by a sequence of states: *putting customer copy of invoice in box, closing box,* and *taping box.*

Since transitions connect states, transition state nets begin and end with a transition. The trigger of the entry transition can always be the trigger of the high-level transition. The trigger in transition *1* in Fig. 5.28(b) is the trigger of transition *10* in Fig. 5.28(a). Exit transitions are often just completion transitions. In Fig. 5.28(b), when a packager finishes taping the box, the high-level transition completes. Since, as a shorthand, we may omit transitions that have no actions and fire only when the actions of a prior state complete, we may omit transition *4* in Fig. 5.28(b). In this case, we still retain the final arrow, which denotes the default transition in this context. The transition identifiers within square brackets on the low-level state net in Fig. 5.28(b) are local in scope and do not have to be unique with respect to transition identifiers in higher level state-net diagrams.

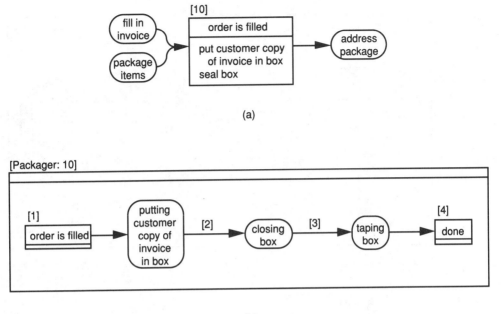

(a)

(b)

Figure 5.28   Top-down state net development for a high-level transition.

If there are several entry transitions, one is chosen, arbitrarily in the absence of any applicable constraints. If there are several exit transitions, an object need only reach one of them unless there are concurrent activities. If there are concurrent activities in a state net for a high-level transition, the transition completes only when all activities finish.

5.4.2 Bottom-Up Development of Transitions

Having discussed top-down development for actions in high-level transitions, we now turn our attention to bottom-up development. Since in bottom-up development all the components of the high-level transition are present, we can automatically derive a state net for a high-level transition. As we discussed in Chap. 4, however, we cannot automatically derive the trigger and action of the high-level transition because the triggers and actions of the transitions in the low-level components are informal.

Figure 5.29 shows an example. We begin with the state net in Fig. 5.29(a), which shows a portion of the packager state net. We create a high-level transition from the cloud-enclosed state and transitions. When we implode the cloud and supply the contents of the high-level transition as discussed in Chap. 4, we obtain the diagram in Fig. 5.29(b). Figure 5.29(c) shows the derived low-level state net for the high-level transition in Fig. 5.29(b).

The diagrams in Figs. 5.28 and 5.29 have simple entry and exit requirements. In both examples there is a single entry and exit. More complex examples require additional information to specify completely the connections for boundary-crossing in- and out-arrows.

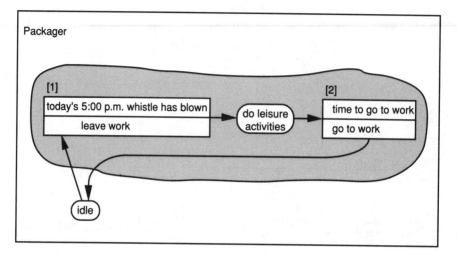

(a)

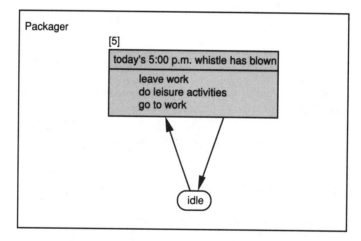

(b)

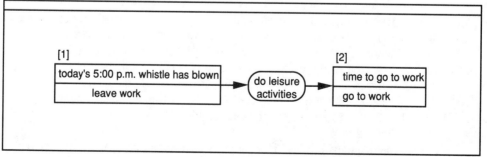

(c)

Figure 5.29   Bottom-up state net development for a high-level transition.

Figure 5.30(a), for example, shows part of a state net with multiple entries into a high-level transition. When we implode the high-level transition, we show the same combinations of enabling states, but we attach the arrow heads to the high-level transition as Fig. 5.30(b) shows. Multiple arrows with identical heads and tails are only represented once. Notice, for example, that only one in-arrow connects state S_1 to the high-level transition even though S_1 is a prior state for two of the transitions inside the high-level transition. Figure 5.30(c) shows how we denote the connections in the state net for the high-level transition. As with high-level states, entry specifications limit the entry. Unlike high-level entry specifications for states, which cannot have multiple prior

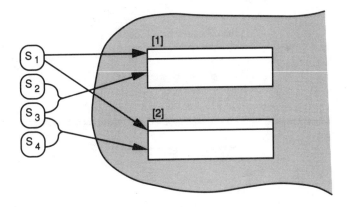

(a)

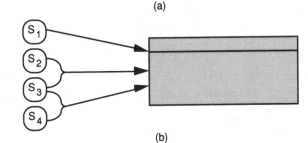

(b)

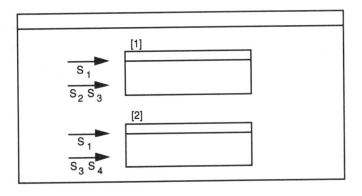

(c)

Figure 5.30   Entry specifications for a high-level transition state net.

transitions, we must specify the combinations of possible enabling states. We do so by specifying a group for each incoming arrow that includes all states connected to its tail. Notationally, we write the group as a list of state names with an arrow above the group and with the arrowhead pointing toward the entry transition to which it applies.

We represent exit transitions and their connection to subsequent states in a similar way. Figure 5.31(a), for example, shows part of a state net with multiple threads of control ending with one of two exit transitions. Figure 5.31(b) shows the imploded high-level transition. Figure 5.31(c) shows the use of exit specifications in a state net for the

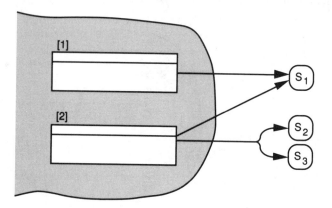

(a)

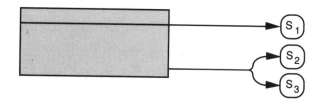

(b)

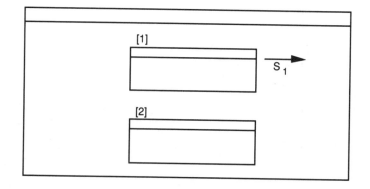

(c)

Figure 5.31   Exit specifications for a high-level transition state net.

imploded high-level transition. The representation for exit specifications is similar to the representation for entry specifications except that the arrow points away from rather than toward the exit transition.

Notice that there is no exit specification for transition *2* in Fig. 5.31(c). In the absence of entry and exit specifications, there are no entry and exit limitations. This means that if an entry transition for some high-level transition has no entry specification, it is assumed that all in-arrows for the high-level transition are in-arrows for the entry transition. Similarly, if an exit transition has no exit specifications, it is assumed that all non-exception out-arrows for the high-level transition are out-arrows for the final transition. For this reason there is no exit specification for transition *2* in Fig. 5.31(c). This is also the reason there are no entry and exit specifications in Figs. 5.28(b) and 5.29(c).

Exceptions entering high-level transitions can be treated as any entering in-arrow, but exceptions leaving high-level transition state nets require some additional notation. Figure 5.32 shows an example. For the exception out-arrow in Fig. 5.32(a), there are two cases to consider: (1) when the exception emanates from a transition within the transition state net, and (2) when the exception emanates from a state within the transition state net. We treat the first case essentially the same as an ordinary exit from a transition state net. The only difference is that we retain the condition and the exception indicator

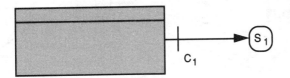

(a)

(b)

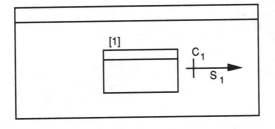

(c)

Figure 5.32 Exit specifications for exceptions in a high-level transition state net.

for the exit specification as Fig. 5.32(b) shows. In the second case, the exception is transformed from a transition exception into a state exception. This transformation requires that the condition for the exception become the trigger for an exit state whose action is empty as Fig. 5.32(c) shows. In this case it is unnecessary to supply exit specifications because the exception condition identifies the out-arrow to which the exit transition attaches.

We show as a final example a state net for a transition in the Green-Grow Seed Company that includes both exception and non-exception exits. Figure 5.33 is a state net for transition *9* of our packager state net in Fig. 5.18. It shows the details of how a packager behaves in transition *9* and how exceptions are detected. The entry transition has the same trigger as transition *9* in Fig. 5.18, but there are no actions. Since a packager initially needs an unfilled order, a box, and an invoice form, the packager may choose arbitrarily which transition to follow out of the entry transition. After getting an unfilled order, box, or invoice form, a packager gets whatever else is still needed. A

[Packager: 9]

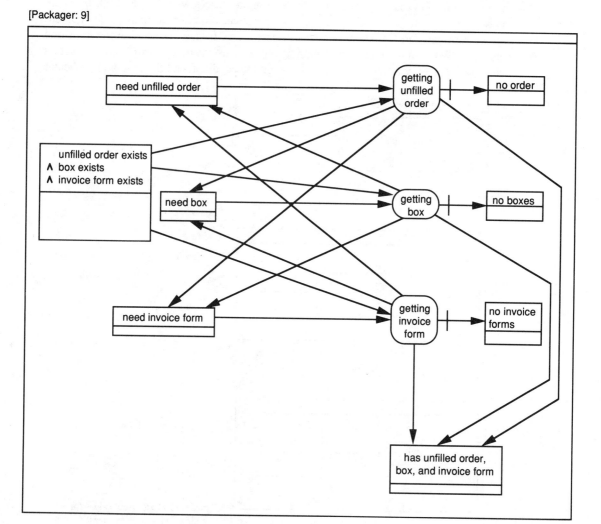

Figure 5.33   State net showing exception handling details.

packager arrives at the non-exception exit transition only if an unfilled order, a box, and an invoice form have been obtained. Otherwise, there is an exception. Either there is no unfilled order at the time a packager attempts to get one, or no box, or no invoice form. Each exit transition for these exceptions includes the exception condition for its trigger.

Observe that, for this example, our rules make it unnecessary to provide limitations on entry or exit specifications. If a packager reaches the non-exception exit transition, the packager exits through the single non-exception out-arrow of high-level transition *9*. Otherwise, a packager reaches a specific exception condition and exits on the out-arrow associated with the exception condition.

5.5 Bibliographic Notes

Much of the material in this chapter is new. Top-down development in general is an old topic, but top-down development for object-oriented analysis models has only recently been discussed [deChampeaux 1991]. The semantics of high-level object classes has some similarities with the notion of a complex object found in object-oriented database literature. Some additional detail on complex objects in OSA can be found in [Czejdo 1991]. High-level relationship-set views are based on relational views from database systems and on work on semantic-model query languages [Czejdo 1990]. Semantics of high-level states and transitions for OSA are new, although high-level states do have some similarities with Harel's statecharts [Harel 1988].

5.6 Exercises

5.1 Consider an automobile loan from the KWE Bank as an independent high-level object class. Object classes internal to the high-level automobile loan are Customer, Loan Amount, Loan Number, Automobile Model, Automobile Serial Number, Monthly Payment, Total Payments, Date of First Payment, and Interest Rate. An automobile loan at KWE Bank may be associated with one or two customers. The possible total number of payments are 24, 36, or 60. Using the top-down approach, develop an ORM for the internal object classes for the high-level Automobile Loan class. Connect the internal object classes to the high-level object class so that it is semantically correct. Use as few connections as possible.

5.2 Figure 5.34 shows a variation of the New Year's Club ORM for the KWE Bank that includes another type of bonus based on ten or more consecutive years as an account customer of the bank. Derive a high-level relationship set between customers and bonuses that includes only those customer-bonus pairs in which the customer qualifies for both a New Year's Club bonus and as Ten-or-More Year bonuses.

5.3 Figure 5.35 shows a partial state net that models the behavior of cash being placed into an ATM's cash drawer opening. The *delivering cash* state is a high-level state. Cash is counted by ejecting the appropriate unit of cash from the cash bin into the drawer-opening chute. Model the low-level behavior of the *delivering cash* state with a state net that exhibits the following behavior. There is a real-time constraint that a unit of cash must pass through the drawer-opening chute in less than 0.5 seconds from the moment the unit of cash was ejected from the cash bin. If this process takes longer than 0.5 seconds, the cash drawer is considered to be jammed.

5.4 Figure 5.36 shows a partial state net for the power-on sequence of the ATM. Transition *1* is a high-level transition. Consider the behavior described by the actions in transition *1*, including the exception conditions and model the low-level behavior concepts of the transition with a state net.

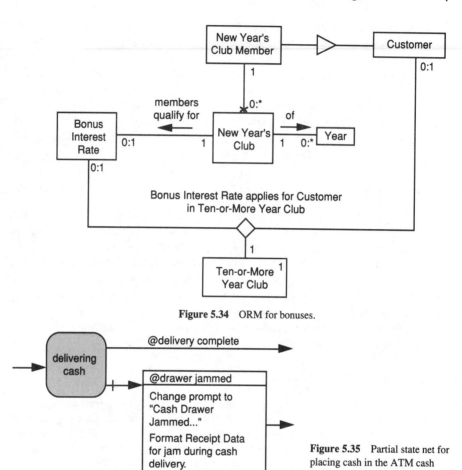

Figure 5.34   ORM for bonuses.

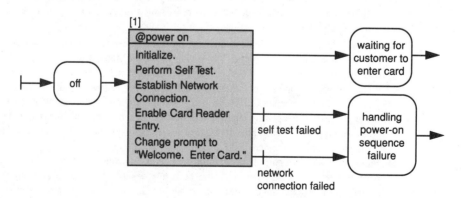

Figure 5.35   Partial state net for placing cash in the ATM cash drawer.

5.5 Model the behavior of the ATM as an aggregate object using the techniques in Sec. 5.3.3. Use behavior scenarios *ATM.13* and *ATM.14* given in the beginning of the exercises for Chap. 3 as the basis for the state-net behavior. Consider *Transaction Printer, Cash Drawer,* and *Card Reader* objects as subparts of the ATM in your state net.

Figure 5.36   Partial state net for ATM power-on sequence.

6

The Object-Interaction Model

The object-relationship model allows an analyst to describe the relationships among objects. The object-behavior model allows an analyst to describe the behavior of an object, but in isolation from other objects. In this chapter we present the object-interaction model, which allows an analyst to describe the interaction among objects.

One object may interact with another in many different ways. For example, an object may send information to another object, an object may request information from another object, an object may alter another object, and an object may cause another object to do some action.

To understand object interaction, we must understand: (1) what objects are involved in the interaction, (2) how the objects act or react in the interaction, and (3) the nature of the interaction. Since we are able to identify objects in ORMs, we use ORM components in object-interaction models to show which objects are involved in interactions. To understand how objects may act or react in a given interaction, we must understand the behavior of each object involved. Since we are able to define the behavior of objects with state nets, we use state nets in object-interaction models to describe how objects act and react in interactions. To understand the nature of an interaction, we must describe the activity that constitutes the interaction, and we must describe the information or objects transmitted or exchanged in the interaction. We thus introduce a feature into OSA that describes the interaction and the objects exchanged in the interaction. We then use this feature with an appropriate combination of ORMs and state nets to create object-interaction models.

We begin our discussion in this chapter with an example that introduces and defines the basic elements of object interaction. After defining basic object interaction, we then continue by discussing specific types of interactions and constraints that can be imposed on interactions. During this discussion, we flesh out the basic definition and further illustrate the use of object interaction in OSA. In defining object interaction, we use our recommended symbols, but again urge the reader to look beyond the symbols to the underlying concepts.

6.1 Basic Interaction Diagrams

Figure 6.1 shows a basic OSA object-interaction diagram used in the Green-Grow Seed Company. *Clerk* and *Packager* are object classes. The zigzag arrow connecting *Clerk* and *Packager* denotes that there is an interaction between a clerk and a packager and that the flow of the interaction is from a clerk to a packager. The label *retrieve (unfilled order)* on the arrow describes the interaction. Thus, the interaction diagram in Fig. 6.1 specifies that there is an interaction between a clerk and a packager that results in a packager retrieving an unfilled order.

In OSA, we have chosen to represent interaction by a zigzag arrow called an *interaction arrow* or an *interaction link*. The flow of an interaction arrow is from an origin at its tail to a destination at its head. The origin is any (single) object in the object class associated with the tail of the interaction arrow, and the destination is any (single) object in the object class associated with the head of the interaction arrow. In Fig. 6.1 a clerk in the *Clerk* object class is an origin for the interaction, and a packager in the *Packager* object class is a destination. The unfilled order flows from a clerk to a packager.

Labels on interaction links describe the interaction and the objects transferred in the interaction. For reasons that will be explained later, we place labels on the arrowhead side of a zigzag. Notationally, we list the descriptions of objects transmitted in the interaction inside parentheses and write a description of the activity outside parentheses. Thus, the activity for the interaction in Fig. 6.1 is *retrieve*, and the object transmitted in the interaction is an *unfilled order*. Objects transmitted in an interaction may be tangible or intangible. In Fig. 6.1, we may consider an *unfilled order* to be tangible (the form with the order information on it) or intangible (the information on an order form). Often, as happens here, it does not matter whether transmitted objects are tangible or intangible.

In the diagram in Fig. 6.1 we have shaded the zigzag part of the interaction arrow. Like high-level object classes, high-level relationship sets, high-level states, and high-level transitions, we also have high-level interaction links. If we wish to know more about the interaction, we can investigate its details. Figure 6.2 shows the details for the high-level interaction link in Fig. 6.1. In Fig. 6.2 we see that an unfilled order is transferred from a clerk to a packager by way of an out-basket. An unfilled order is first transferred from the *Clerk* to the *Clerk Out-Basket*. This takes place during the *deposit*

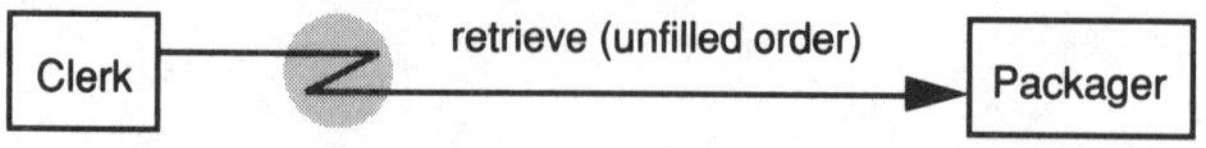

Figure 6.1   Basic object interaction.

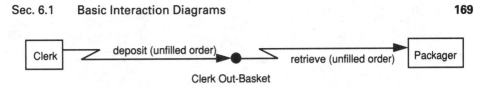

Figure 6.2   Interaction diagram with an intermediate repository.

interaction. The unfilled order is then transferred from the *Clerk Out-Basket* to the *Packager*. This takes place during the *retrieve* interaction.

To find out even more about the interaction, we can investigate the behavior of clerks and packagers by looking at their state nets. Figure 6.3 shows an example. Here we have two transitions, one that is part of a state net for clerks and one that is part of a state net for packagers. Indeed, the transition for the packager state net is transition *9* of our packager state net developed in Chap. 3. Figure 6.3 shows that using state net information we can describe the actions taking place during an interaction, what triggered the actions, which objects are performing the actions, and which objects are being acted upon. For instance, during the deposit interaction, a clerk places an unfilled order in the

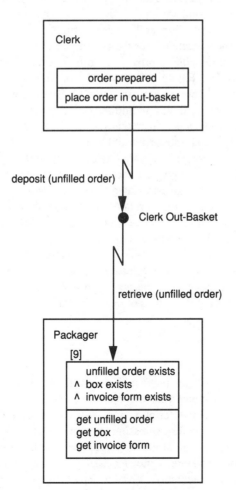

Figure 6.3   Interaction diagram with partial state nets.

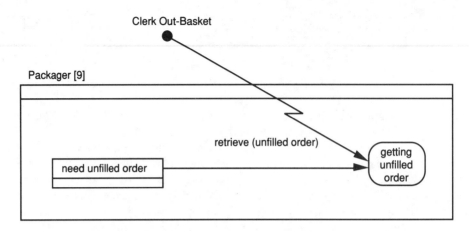

Figure 6.4   Lower level interaction diagram.

out-basket. This interaction is triggered by *order prepared*. During the *retrieve* interaction, a packager gets an unfilled order from the out-basket. This interaction is triggered by the existence of an unfilled order, box, and invoice form.

In Chap. 5, we developed a state net for the action part of transition *9* in the packager state net. We are, therefore, able to provide even more details about when a packager picks up an unfilled order. Figure 6.4 shows part of the state net we developed earlier as a substate net for the action part of transition *9*. By examining the state net in Fig. 6.4, we see that a packager retrieves an unfilled order while in the *getting unfilled order* state, which is entered when a packager needs an unfilled order.

Figures 6.3 and 6.4 show that interaction links may be incident on states and transitions as well as on object classes. Since a state net is associated with an object class, we are able to identify the origin and destination objects as members of the object classes with which the state nets are associated. Hence, when we use state nets in object-interaction diagrams, we are able to retain information about origin and destination objects and gain the additional advantage of being able to identify the specific behavior of the objects involved in the interaction.

In this initial example, we have described the basic features of OSA object interaction. We have also shown how to describe interaction at various levels of abstraction and how to use both object classes and state nets with interaction links. We now continue our discussion by describing the details of object interaction.

6.2 Interaction Descriptions

Figure 6.5 shows the basic formats for interaction descriptions. As can be seen, we may have both an activity description and a list of objects, just an activity description, just a list of objects, or neither an activity description nor a list of objects. Each one of the formats in Fig. 6.5 is a valid technique for expressing interactions among objects.

Figure 6.6 gives an example of an interaction description that has both an activity description and a list of objects. Here, a traveler asks a travel agent to make an airline reservation and, as part of the interaction, gives a departure city, a destination city, a

activity (list of objects)

activity

(list of objects)

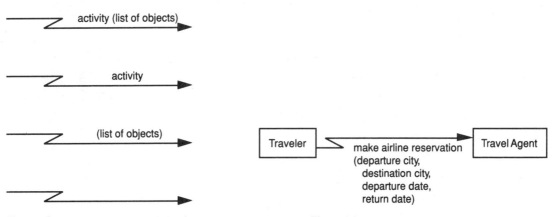

Figure 6.5   Basic interaction-description formats.   **Figure 6.6**   Full interaction description.

departure date, and a return date. Unlike our initial example in which only one object is transmitted in an interaction, this example shows that several objects may be transmitted in an interaction. When a list of objects is transmitted as in Fig. 6.6, the destination object receives them all during the interaction.

Figure 6.7 gives an example of an interaction description that has only an activity description. Here, a driver interacts with a vehicle by driving it. No objects are transmitted in this interaction.

The most common type of interaction among objects is communication. Hence, we choose to make communication our default activity description. Two objects communicate when one object sends a message to another object. Since messages are tangible or intangible objects transmitted in an interaction, we represent them in interaction diagrams within parentheses in interaction labels. The activity description is *communicate,* or any of numerous possible synonyms, but is not written since it is the default. In this context we sometimes refer to an interaction link (interaction arrow) as a communication link (communication arrow). Figure 6.8 shows an example of a communication link. A drowning person sends the message "*help!*" to a lifeguard.

Occasionally, it is useful to place an interaction link on a diagram without labeling it. Sometimes enough context information is available to make an activity description redundant. In Fig. 6.9, for example, we would be able to infer properly the interaction description without the labels. Because of the activity in the clerk transition, we would know that a clerk places an unfilled order in the out-basket. Knowing that unfilled orders are in the clerk out-basket, we would know that a packager retrieves an unfilled order out of the out-basket. We may even wish to use unlabeled interaction links when there is not enough context information or even when there is no context information. In

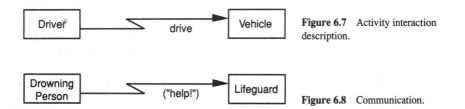

Figure 6.7   Activity interaction description.

Figure 6.8   Communication.

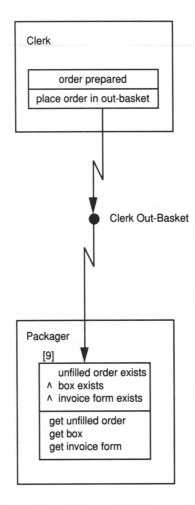

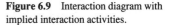

Figure 6.9 Interaction diagram with implied interaction activities.

this case, the interaction link merely indicates that there is an interaction. If we are not interested in showing the details of the interaction in a diagram, we may omit the description.

6.3 Synchronous Object Interaction

It is often necessary to synchronize interacting objects. To successfully communicate in ordinary conversation, for example, the receiver must be ready to receive a message when the sender sends it. If the receiver is not ready, the message is lost. Usually, the sender and receiver understand that certain conditions must be met so that communication can take place, and they see to it that these conditions are met. For example, when two people wish to talk on the telephone, they first establish a connection. If the caller says something before the connection is established, whatever is said is lost. Statements made after the connection is broken are also lost.

Synchronization in OSA is not automatic, but we can use state nets to establish

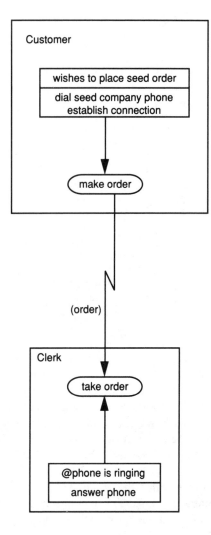

Figure 6.10   Synchronous interaction.

any required synchronization. As an example, Fig. 6.10 shows how a customer places a phone order to a clerk in the Green-Grow Seed Company. We see in the customer state net in Fig. 6.10 that when a customer wants to place a phone order, the customer calls the company and establishes a connection. The clerk, on the other hand, answers the phone when it rings. When a customer is in the *make order* state and a clerk is in the *take order* state, the customer can send a message describing the order, and the clerk can receive the order.

6.4 Asynchronous Object Interaction

Synchronous interaction is not always necessary. For example, people frequently interact indirectly with each other. They may communicate through the mail or through intermediaries. They may also leave objects in a designated place where someone else

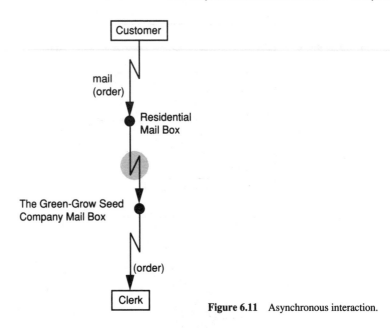

Figure 6.11   Asynchronous interaction.

can pick them up. In our introductory example, for instance, we modeled a clerk depositing an unfilled order in an out-basket, to be picked up by a packager.

In OSA we model asynchronous interaction by providing an intermediate repository such as an out-basket. Since repositories are objects, we can model them using ORMs. Figure 6.11, for example, shows how a customer places a mail order with the Green-Grow Seed Company. There are two intermediate repositories in Fig. 6.11, *Residential Mail Box* and *The Green-Grow Seed Company Mail Box*. A customer mails an order by placing it in a residential mail box. The order then goes through some unspecified additional repositories within the postal system represented by the shaded zigzag on the intermediate interaction link. The mail eventually arrives at the mail box for the seed company. A clerk obtains the order from the company's mail box.

6.5 Specifying Particular Interacting Objects

In all our previous object-interaction examples, any origin object could interact with any destination object. The diagram in Fig. 6.2 allows any clerk to deposit an unfilled order, and any packager to select one to be filled. Similarly, the diagrams in Figs. 6.6 through 6.11 allow any traveler, any travel agent, any driver, any vehicle, any customer, and any clerk to participate in the interaction.

Sometimes, however, we want specific objects to be involved in an interaction. Consider, for example, sending a package containing the seeds ordered back to the customer who ordered them. Here, we want a specific customer to receive the package, not just any customer. An interaction diagram with an interaction arrow from object class *Packager* to object class *Customer* with the label (*package*) alone will not describe the interaction we want.

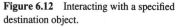

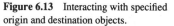

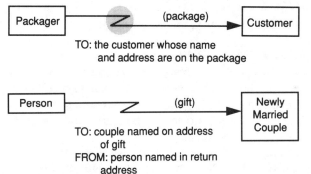

Figure 6.12   Interacting with a specified destination object.

Figure 6.13   Interacting with specified origin and destination objects.

In a real-life situation, we would send a package to a particular customer by addressing it to the customer. In OSA, we do the same. Figure 6.12 shows how. We attach identifying information to an interaction label to direct it to a particular object in an object class. We choose to represent the identifying information in a *TO-clause* that consists of the marker "TO:" followed by the identifying information. In Fig. 6.12 the identifying information in the TO-clause is

the customer whose name and address are on the package.

We are thus able to send the package to the right customer.

Besides identifying the destination object, we may also wish to identify the origin object. We use a FROM-clause to identify an origin object. Similar to a TO-clause, a *FROM-clause* consists of the marker "FROM:" followed by identifying information. Figure 6.13 shows an example. Here a person sends a gift to a newly married couple. The TO-clause identifies the couple receiving the gift, and the FROM-clause identifies the person sending the gift.

6.6 Interacting with Multiple Objects

So far, we have only discussed single origin and destination objects, but interactions can involve multiple origin and destination objects. Broadcasting the news from a radio station to all listeners is an example of a single origin object interacting with multiple destination objects. A group of people moving a piano is an example of multiple origin objects acting on a single destination object. As an example of multiple origin and multiple destination objects interacting, we may consider a chorus singing for an audience.

In OSA, we choose to show a single origin object interacting with multiple destination objects by an interaction arrow with multiple arrowheads. Figure 6.14 shows an example. Here, the manager of the Green-Grow Seed Company sends the message,

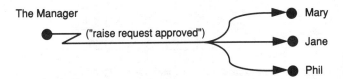

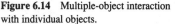

Figure 6.14   Multiple-object interaction with individual objects.

"*raise request approved*" to employees Mary, Jane, and Phil. We may imagine, for example, that the manager calls Mary, Jane, and Phil into the office and tells all three the good news at the same time.

In Fig. 6.14, the arrowheads point at specific objects. Sometimes, we wish to interact with all members of an object class. To show that all members of an object class can be destination objects, we choose to use an interaction link having two arrowheads with intervening ellipses pointing at the object class. Figure 6.15 shows an example. The manager of the Green-Grow Seed Company sends all clerks the message, "*From henceforth, please use only the new order form.*"

When we wish to show multiple origin objects in OSA, we either use multiple tails for individual objects or a pair of tails with intervening ellipses for objects in object classes. Figure 6.16 shows an example. Here, the clerks of the Green-Grow Seed Company are rearranging the office furniture. At this level of abstraction, we are thinking about all the clerks working together to rearrange all the furniture.

We can broaden the scope of multiple interactions to a superset of objects in several object classes by introducing a generalization. Figure 6.17 shows how the manager of the Green-Grow Seed Company can broadcast a message to all clerks and packagers. The communication link goes from the manager to a generalization of object classes *Clerk* and *Packager* called *Hourly Employee*.

We can limit the scope of multiple interactions to a subset of the objects within an object class in two ways. One way is to alter an ORM by introducing a specialization.

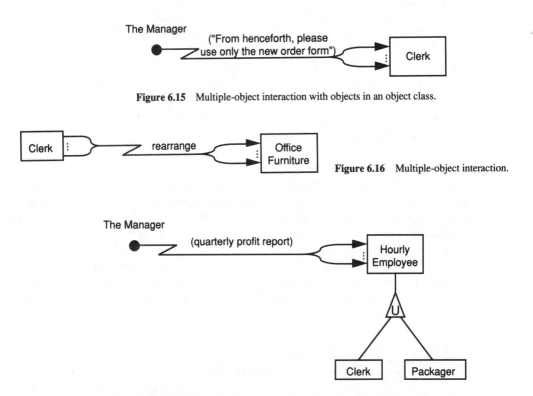

Figure 6.15   Multiple-object interaction with objects in an object class.

Figure 6.16   Multiple-object interaction.

Figure 6.17   Multiple-object interaction with objects in a generalization object class.

The Manager

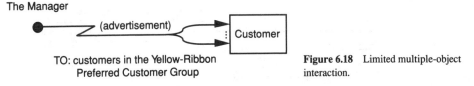

TO: customers in the Yellow-Ribbon
Preferred Customer Group

Figure 6.18   Limited multiple-object
interaction.

Once a specialization is created that includes only the desired subset, we can use it directly. A second way to interact with a subset of objects in an object class is by using TO- and FROM-clauses. Figure 6.18 shows an example. Here, the manager of the Green-Grow Seed Company wishes to send an advertisement to customers, but only to those who are in the Yellow-Ribbon preferred customer group. As the example in Fig. 6.18 shows, we use a restricting phrase in a TO- or FROM-clause that describes the subset of objects involved in a multiple-object interaction.

6.7 Bidirectional Interaction

Sometimes pairs of interactions are so closely related that we want to think of them as being the same interaction. Consider, for example, a bank customer withdrawing cash from an account. Figure 6.19(a) shows the withdrawal as two separate interactions. In the first interaction, a bank customer makes a request to withdraw a certain amount of money from an account identified by an account number. The second interaction shows that a bank teller returns cash to the bank customer requesting the withdrawal. Figure 6.19(b) shows the withdrawal as a single, high-level interaction. The meaning is the same, but the diagram is simpler and there is an implication that we perceive the interaction as a single unit.

We shade the zigzag to show that the interaction is high-level and that we can thus examine its details, which are in the original interaction diagram in Fig. 6.19(a). We may also omit the shading if we do not wish to suggest that lower level details exist. Figure 6.19(c) shows the bidirectional interaction without a shaded zigzag.

Placement of the labels on bidirectional interaction links makes a difference in how we interpret the interaction. If we were to place the *withdraw (account#, amount)* in the middle of the bidirectional interaction arrow in Fig. 6.19(b), we would not know whether the bank customer is requesting the withdrawal of the bank teller or whether the bank teller is requesting the withdrawal of the bank customer. As we have said earlier and as we have illustrated from the beginning, we place the label on the destination-object side. Thus, we know that the origin object for *withdraw (account#, amount)* is a bank customer and that the destination object is a bank teller.

Sometimes, the interaction descriptions on both ends of a bidirectional are identical. In this case, we write only one description and place it in the middle. Figure 6.20 shows an example. Here, a seller negotiates the price of an item with a buyer and vice versa. It would be misleading to put *negotiate (price)* on one end or the other, for then either the seller would negotiate the price with the buyer or the buyer would negotiate the price with the seller, but not both. Although we could properly place the label on both ends, we can avoid this redundancy by placing the label in the middle as Fig. 6.20 shows.

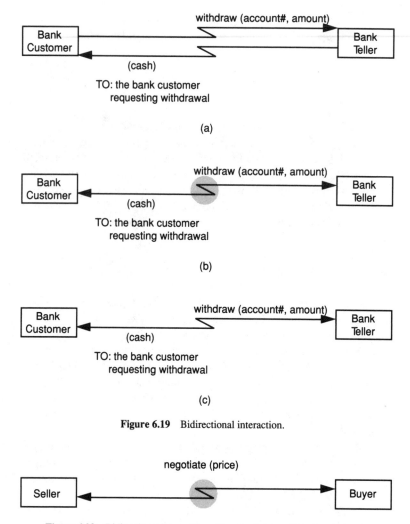

Figure 6.19   Bidirectional interaction.

Figure 6.20   Bidirectional interaction with a single interaction description.

6.8 Special Interaction Activities

The interaction activities—"access," "modify," "remove," "destroy," "add," and "create"—are of particular interest because they are commonly needed for analysis and because we often think of them in a special way. The reason we often view these activities as special is because they can be thought of as an interaction with the system model we are building.

Although we have given these interaction activities particular names for our discussion, the names are not special keywords. It is the concept that is of interest, not the name. Different names may be chosen by the analyst, and the choice usually depends on the context of the analysis. "Create," for example, might be called "hire" for new employees or "matriculate" for entering college freshmen.

6.8.1 Access

Access interactions obtain information about objects. Figure 6.21 illustrates four ways to depict the access of information. The example presumes that while cutting an order, a clerk may need to know the discount rate for a customer.

Figure 6.21(a) shows that when a customer discount rate is needed, a clerk starts the interaction by sending the request *access discount rate* to an object in the high-level *Customer Discount Information* object class. Because of the clause *TO: customer discount information of customer,* the request goes specifically to the high-level customer-discount-information object that contains the information. The object responds by returning the discount rate as a message *(discount rate)* as Fig. 6.21(a) shows.

Figure 6.21(b) shows the same interaction, but this time with some shorthand. First, we make use of a bidirectional interaction link. Second, we shorten *access discount rate* to just *access.* We can do this because "access" means "return what is requested." Since it is redundant to write the information to be accessed both in the request and also in the list of information to be returned, we write it only in the return list. Third, except in special cases such as a singleton object class, "access" requires some information to identify which objects are to be accessed to obtain the requested information. We can put this in a TO-clause as we do in Fig. 6.21(a), but in the context of "access" it makes sense to supply the identifying information as part of the request. We thus omit the TO-clause and append an object list to the activity name as Fig. 6.21(b) shows.

Figure 6.21(c) again shows the same interaction, but this time with a different destination object. Here, *The System* is the analysis system, which is a singleton high-level object class that includes all object classes and relationship sets contained in the model. The system, therefore, contains, in particular, information about customers and customer discount rates. There is no conceptual difference between accessing information included within a high-level object belonging to a specified high-level object class such as *Customer Discount Information* and accessing information in *The System* object, which is a high-level object that includes all information. During design when we are interested in saying how an access is made, we are likely to prefer the more-limited version in Fig. 6.21(a), but during analysis we should make our choice based on the best way to document the concept and avoid premature decisions regarding implementation details.

Figure 6.21(d) shows one more way to document the same interaction. In this version, there is no interaction arrow. We know that there is an interaction, however, because of our understanding of the activity description in the action part of the transition in the clerk state net. The action listed is *get discount rate for customer.* If this is sufficient, and here it may well be, we need not document the interaction using an interaction arrow.

6.8.2 Modify

Modify interactions alter existing objects. As an example, Fig. 6.22 shows two ways we can depict a clerk altering an incorrect social security number. In Fig. 6.22(a), we show a full version without shorthand. The clerk may be thought of as directly modifying the incorrect social security number. The *SSN* to be modified is in the clause *TO: the incorrect SSN* and the new *SSN* is part of the interaction description.

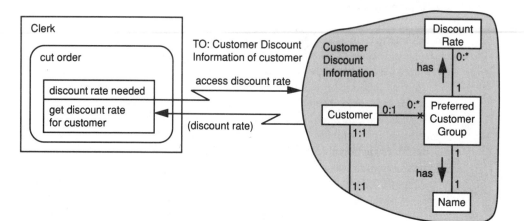

(a)

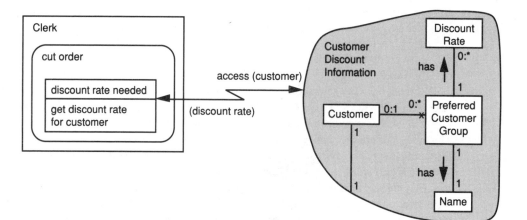

(b)

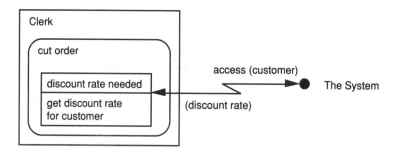

(c)

Figure 6.21   Four ways to do access.

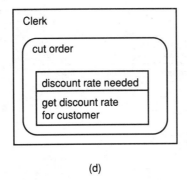

(d)

Figure 6.21 (Cont'd.)   Four ways to do access.

Figure 6.22(b) shows a shorthand for this interaction. Since we are modifying an object in the destination object class, we need to identify it and provide the modification. Here, we identify it as the *incorrect SSN* and provide the modification as a replacement object called the *new SSN*.

We may also replace the destination object class *SSN* by the singleton object class *The System* as we did in Fig. 6.21(c), and we may omit the interaction altogether as we did in Fig. 6.21(d). In both cases, the conceptual meaning would be the same.

In Fig. 6.22 the *SSN* object class is atomic. We may also modify high-level objects. Figure 6.23 shows a clerk modifying the name of a person. Here, *Person* is a dominant high-level object class. We can thus identify an object in the object class as a person. The interaction description *modify (person, new name)* thus uses *person* to identify the object to be modified and uses *new name* to make the modification. As a result

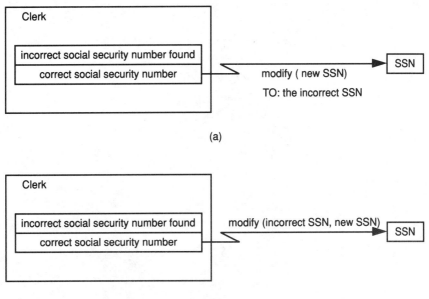

Figure 6.22   Object modification.

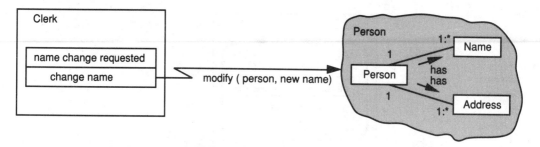

Figure 6.23 Modification of objects and relationships in a high-level object class.

of this modification, we may assume that a relationship between the identified person and the new name is established and that all the constraints of the model are satisfied. During analysis, we need not concern ourselves about how the modification is done as we would during design or implementation. Presumably, in this case, the system would sever the link between the identified person and the old name, remove the old name from *Name* if it is no longer the name of any person, search for the new name, insert the new name if it does not already exist, and establish the relationship between the person and the new name.

6.8.3 Remove and Destroy

Remove deletes an object from an object class, but not necessarily from the analysis model. Figure 6.24 shows an example. When the manager terminates an employee, the employee is removed from the records of the Green-Grow Seed Company. The

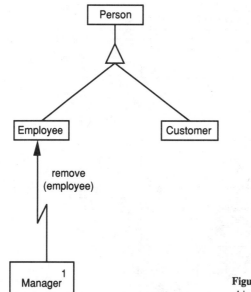

Figure 6.24 Deletion of an object from an object class but not from the model.

employee, however, is still a person and may still be a customer. We, therefore, do not wish to remove the employee from either the *Person* object class or the *Customer* object class. In Fig. 6.24, we only show the shorthand version for removal. In a full version, we would have a TO-clause that identifies the employee object participating in the interaction.

While doing analysis, we would not normally concern ourselves about the propagation required within the model to remove an object from an object class. In our full seed-company example, for instance, the terminated employee's social security number and the relationship instance between the terminated employee and social security number as well as several other objects and relationships would also need to be deleted. We should assume, without having to specify anything, that all additional deletions take place properly.

As part of propagating deletions correctly, we must consider that the object is removed from all specializations. Thus, for example, when the manager terminates an employee who is a clerk, then the employee is also no longer a clerk. Removal, however, does not propagate to generalizations. Thus, a former employee may still be in object class *Person,* and, of course, may still also be in object class *Customer.*

If we want to propagate to generalizations and thus delete all knowledge of an object from the model, we would destroy the object rather than merely remove it. Destroy causes an existing object to cease to exist as far as the analysis model is concerned. In Fig. 6.24, if we replace *remove* with *destroy,* the employee would cease to exist (as far as the model is concerned) even as a person and a customer.

6.8.4 Create and Add

Create brings an object into existence within the analysis system being modeled. Suppose, for example, that we wish to show that a clerk creates an order. In Fig. 6.25 we show the *cut order* state, which is part of the behavior of a clerk for the Green-Grow Seed Company. The interaction link in Fig. 6.25 from the *cut order* state to object class *Order* specifies that a clerk creates a new order.

The interaction diagram in Fig. 6.25 looks simple enough, but it takes some careful thought to reconcile it with our notion of interaction. To begin with, we observe that before the interaction takes place, there is no destination object. In our example, the order does not exist until the clerk creates it. We may reconcile this point by realizing that the clerk is interacting with the order being created and that during the time the interaction takes place, the order becomes part of the *Order* object class. Even though we can reconcile the initial absence of the interacting object, we would probably find it strange to use a TO-clause to specifically identify the object before it is created. Indeed, we cannot identify it until it is created. We, therefore, depict creation only in the shorthand as Fig. 6.25 shows.

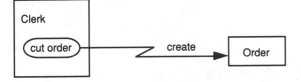

Figure 6.25   Object creation.

A second problem is that we normally cannot just put an object in an object class without also providing some additional information. An order, for example, requires an order#. It is possible to resolve this problem by using a high-level object class that includes all required object classes and relationship sets and by supplying the objects that either need to be added to or referenced in these object classes. Figure 6.26 gives an example of what would be required for creating an order for the Green-Grow Seed Company. During analysis, however, we may not wish to supply all this detail or even concern ourselves with the possible extent of the propagation. We thus allow interaction diagrams, such as the one in Fig. 6.25, where the information required to successfully complete an interaction involving creation is implicit. We can determine what the information is by examining the destination object class in its ORM context.

A third problem is that we cannot simply "plug in" the singleton object class *The System* in place of a more-specific destination object for the create activity. The reason

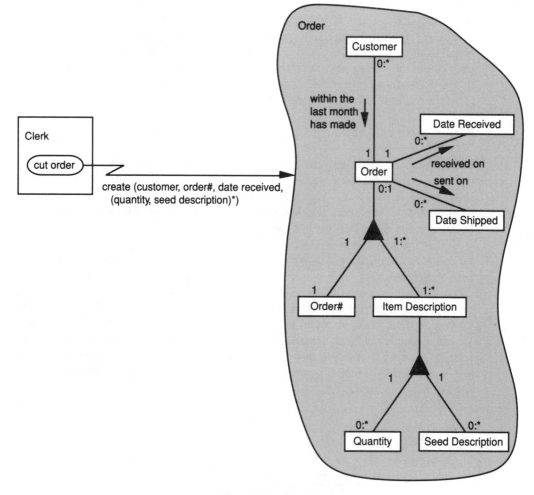

Figure 6.26   Complete creation.

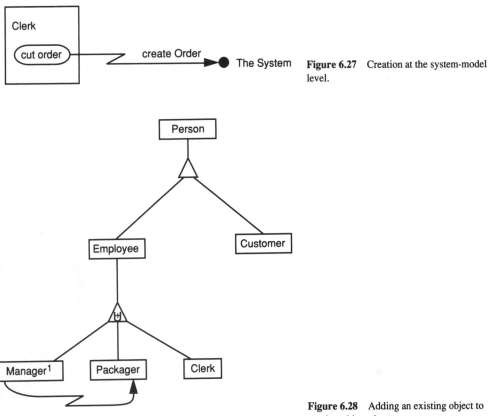

Figure 6.27   Creation at the system-model level.

Figure 6.28   Adding an existing object to another object class.

is that there is no information that identifies the destination object class when create alone is used as the interaction description. By supplying the destination object class, we can solve this problem as Fig. 6.27 shows. Thus, the interaction description is *create Order* rather than *create* alone.

Add is similar to create because it adds an object to an object class, but differs because the object to be added already exists in the model. Suppose, for example, that we hire a packager for the Green-Grow Seed Company who is already a customer of the company. Figure 6.28 shows that the manager adds a new packager by giving identifying information for the person to be added as a packager. Adding has the same propagation problems as creating, but the solutions are also the same.

6.9 Bulletin-Board Communication

One common way for objects to communicate is through a "bulletin-board" protocol. One object posts a message, and others read it. We can use the ideas just discussed for create, destroy, and access to model bulletin-board communication.

Figure 6.29 shows an example of bulletin-board communication with object classes *Person, Secretary,* and *Message.* The interaction links in Fig. 6.29 show that

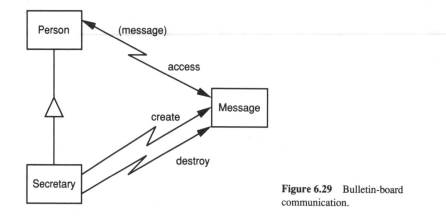

Figure 6.29   Bulletin-board communication.

secretaries can create messages for the object class *Message*. They can also destroy messages. A person can access all the messages in the *Message* object class and can therefore read all current messages posted by a secretary. This modeling technique may be used for any conceptual or concrete instance of bulletin-board communication found in a system.

6.10 Model Boundary-Crossing Interactions

We occasionally wish to show interactions whose origin or destination is beyond the boundaries of the analysis model, and therefore, not known. We may, for example, wish to declare that a message or an object comes into the model from an unspecified origin. We may also wish to declare that a message or object goes out of the model to an unspecified destination.

We choose to represent unspecified origins and destinations by their absence at the tail(s) or head(s) of an interaction link. Figure 6.30 shows an example. The manager of the Green-Grow Seed Company receives boxes and seed packages from unspecified origins and sends advertisements to unspecified destinations. The advertisements are broadcast to anyone who will receive them.

Figure 6.30   Model boundary-crossing interactions.

6.11 Continuous Interaction

Sometimes an interaction is continuous. This is common particularly for analogue sensors in systems. A heat sensor on a boiler, for example, continuously checks and emits the current temperature. Similarly, a speed sensor in a cruise-control system continuously checks and emits the current speed.

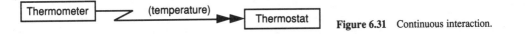

Figure 6.31   Continuous interaction.

If a destination object continuously checks information, the information must be continuously available from the origin object. On the other hand, even if an origin object continuously sends information, a destination object does not necessarily continuously receive information. A maintenance person responsible for a boiler, for example, only gets the temperature from the heat sensor during boiler maintenance checks. Although a wrist watch continuously tells time, a person wearing the wrist watch does not continuously check the time. Destination objects may miss information available to them, but it often does not matter. What often does matter is the continuous availability of information.

We have chosen to show that an interaction is continuous with a double arrowhead. Figure 6.31 shows that a thermometer can continuously supply the temperature for a thermostat. In this example the thermostat, which is also an analogue device, just happens to also be able to continuously receive the temperature.

6.12 Time-Constrained Interactions

If an interaction does not take place within a certain time, it may have no value or may have greatly diminished value. We may, therefore, wish to constrain the time required for an interaction to complete. We say that an interaction completes if all origin and destination objects participating in the interaction fulfill their role in the interaction. An interaction does not complete, for example, if an origin object sends a message to a destination object, but the destination object never receives it, or if an origin object sends a message to multiple destination objects, but not all receive it.

In OSA, we declare time-constrained interaction using real-time constraints. We use the same notation for real-time constraints introduced in Chap. 3. When we place a real-time constraint on an interaction link, we declare that the interaction should complete within a specified time. Figure 6.32 for example, shows that an excessive heat warning should be sent from a nuclear reactor and received by a controller in less than 10 milli-seconds.

It may also be desirable or necessary to place time constraints on sequences of interactions. To declare real-time constraints for interaction sequences, we place real-time markers on interactions paths and then constrain the time from marker to marker. In Fig. 6.33, for example, we show how to constrain the time it takes for a customer to receive seeds ordered from the Green-Grow Seed Company. We place the marker {a} near the tail of the communication arrow that describes the order being placed, and the

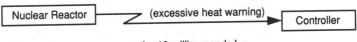

{ < 10 milli-seconds }

Figure 6.32   A real-time constraint on an interaction link.

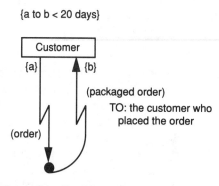

{a to b < 20 days}

The Green-Grow Seed Company

Figure 6.33   A real-time constraint on an interaction path.

marker $\{b\}$ near the head of the communication arrow that describes the order being shipped. The real-time constraint

$$\{a \text{ to } b < 20 \text{ days}\}$$

constrains the time it takes from making an order to receiving the packaged order to be less than 20 days.

6.13 General Interaction Constraints

In our last example in Fig. 6.33, there is an unstated implication that the order placed corresponds to the packaged order received. To say so explicitly, we can use a general constraint. We may use natural language to express our constraints and thus to ensure the correspondence, we might write the following constraint on the interaction diagram:

The order placed at time {a} corresponds to the order received at time {b}.

We may use general constraints in a variety of ways to address several interaction issues. In communication models, for example, we may constrain communication rates, bandwidth, quality of communication, and limitations on when messages can be sent and received. For asynchronous interaction, we can place priority constraints on which objects should have highest priority in the interaction. As an example, Fig. 6.34 shows

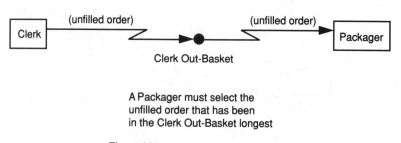

A Packager must select the
unfilled order that has been
in the Clerk Out-Basket longest

Figure 6.34   Interaction with a constraint.

how to ensure that packagers process unfilled orders on a first-in/first-out basis by requiring that packagers select the unfilled order that has been in the out-basket longest.

6.14 Interaction Within an Object Class

Sometimes we may need to model communication among objects in the same class and even objects communicating with themselves. People, for example, often leave messages for themselves, and, despite the negative connotation, they also talk to themselves. They also leave objects in various places intending to pick them up later, and they also carry objects with them when they go from one task to another. In OSA, we show that an object interacts with itself by using interaction links that go from a state or transition of a state net to another state or transition in that object's state net.

Figure 6.35 shows an example of an object interacting with itself. When a packager exits transition *9,* the packager takes along an unfilled order, a box, and an invoice form. While a packager is packaging items and filling in the invoice form, information about the number of packages placed in the box flows from the activity in the *package items* state to the activity in the *fill in invoice* state. When a packager finishes filling an

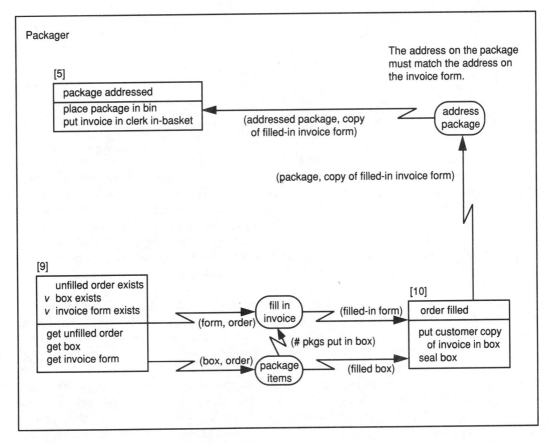

Figure 6.35   Interaction within an object.

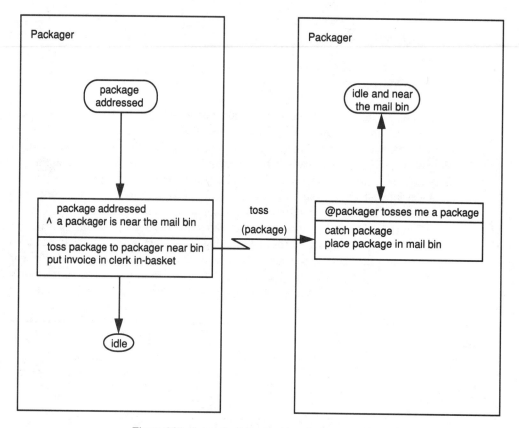

Figure 6.36   Interaction between objects in the same object class.

order, the box with seed packages in it and the filled-in invoice are taken to transition *10* in which the packager places the customer copy of the filled-in invoice in the box and seals the box. The packager next takes the sealed box, which is now called a package, and the clerk copy of the filled-in invoice to the *address package* state. When a packager finishes addressing a package, the package, along with the clerk copy of the filled-in invoice are taken into transition *5*. The package address, by the way, must match the address on the invoice form as required by the general constraint in Fig. 6.35. In transition *5* the packager places the package in the mail bin and the clerk copy of the invoice form in the clerk in-basket.

Even without the interaction links, many of the interactions in Fig. 6.35 would be well understood. As illustrated previously in Fig. 6.21(d), there is no requirement that interactions be explicitly documented using interaction links. One reasonable alternative for Fig. 6.35 is to leave out all the interaction links except the one between the states *fill in invoice* and *package items*.

Besides interaction within an object, we may also wish to show interaction among different objects in the same object class. All objects in an object class share the same state-net template, but each object has its own copy. We may, therefore, show interactions between two different objects in the same object class by showing interactions between two copies of the object class's state-net template.

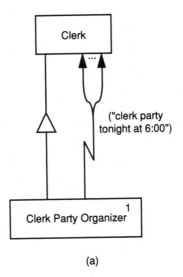

(a)

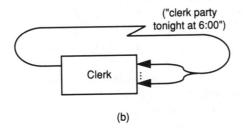

(b)

Figure 6.37 Interaction among objects in the same object class.

Figure 6.36 shows an example of two packagers interacting. The state nets on the left and right in Fig. 6.36 are both partial state nets for packagers and both show behavior beyond the behavior we have presented earlier for packagers. The state net on the left shows that after a packager has addressed a package, the packager may toss it to another packager standing near the mail bin rather than carry it to the mail bin. The state net on the right shows that if a packager is idle, is standing near the mail bin, and another packager tosses a package, the packager catches it and places it in the mail bin. The interaction arrow shows that one packager can toss a package to another packager.

We can also show interaction within an object class at the ORM level. Figure 6.37(a) shows that a party organizer, who is also a clerk, can broadcast a message, *"party tonight at 6:00,"* to all clerks. In Fig. 6.37(b) we have collapsed the *Clerk Party Organizer* object back into object class *Clerk* so that it cannot be seen. We thus show that some clerk, rather than a particular clerk, announces the party. When a communication link is self-referencing with respect to an object class, it declares the possibility of an interaction between or among objects of the same object class.

6.15 High-Level Interactions

Specification of object interaction can be involved and complex. Often, however, we are interested only in a few pertinent facts. Using the principle of abstraction, we can express only what is needed by omitting some of the detail. With nonessential detail omitted, we can focus more easily on pertinent ideas.

We have used this principle of abstraction throughout this chapter. For example, in Fig. 6.1 we showed the interaction between a clerk and a packager without giving the details about the clerk out-basket as an intermediate repository. In Fig. 6.3 we showed partial state nets for a clerk and a packager that just included one transition each. For Fig. 6.5 we explained that we use communication as the default activity for an interaction and also that interaction arrows without any description are useful when we only need to indicate that there is an interaction. In Figs. 6.19 and 6.20 we showed how to use bidirectional arrows to simplify pairs of related interactions. In Fig. 6.21 we not only showed the general forms of interaction for special activities, but we also showed several ways to more succinctly express the same interaction, including the omission of interaction links altogether. In this section we discuss a few more high-level interaction concepts.

6.15.1 Interaction Sequences

It is common to have several interactions between the same two objects. To show these interactions, we can place the state nets of the object classes of the two objects in a diagram, with all the interactions and the sequence in which we expect the interactions to occur. The detail, however, may be overwhelming. Furthermore, we may not have developed the state nets for the object classes, and we may not want to interrupt our train of thought about the interactions by stopping to develop the state nets.

As an alternative to showing interactions at the state-net level, we can abstract up to the object-class level and just show the interactions without the state nets. Figure 6.38, for example, shows several interactions between a bank teller and a customer's bank account. We assume that a bank customer has requested the withdrawal of a certain amount of money from an account. Before the teller returns the money to the customer, the teller must verify several things to be sure that the customer should receive the money. The account number must be valid, the customer must be a co-signer on the account, and the account must not be overdrawn. We show these interactions in Fig. 6.38.

Observe that we have taken several liberties with our notation. First, the teller clearly interacts with a specific bank account, namely the account whose account number is part of the activity description in the top interaction arrow in Fig. 6.38. Since this is clear from context, we omit the clause *TO: the bank account designated by account#*. Second, the *validate signer name* interaction and the *withdraw (amount)* interaction also interact with a specific account, but here we do not even supply the account number nor do we use a TO-clause. Again, context makes it clear that we are interacting with the same bank account. Clearly, the same teller also participates in all the interactions. Strictly speaking, we would not know that these interactions have anything to do with one another taken individually, but again, because of context we view them as a unit.

By considering the interactions in Fig. 6.38 as a coordinated unit, we gain an unexpected benefit. The group of interactions begin to give us a glimpse of the (as yet) unspecified state nets. Indeed, one approach to the development of state nets is to first

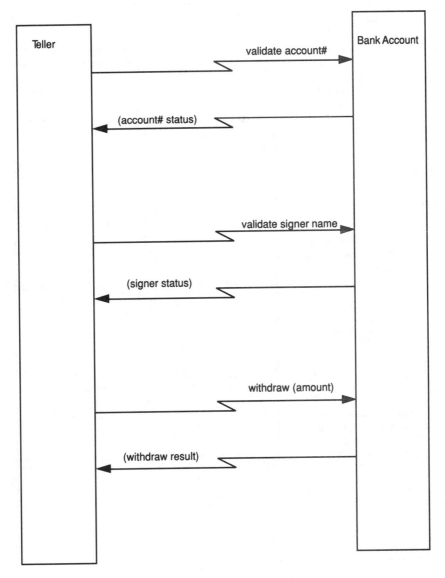

Figure 6.38   A sequence of interactions.

consider the interactions of an object in an object class, and then piece together a behavior that satisfies the interactions.

Figure 6.39 shows how we might begin to create a state net in this fashion. Here, we place some natural-language description of a teller's behavior inside the *Teller* object class. Figure 6.40 carries this one step further by providing some of the components of the eventual state net. It also illustrates yet another type of high-level diagram, one where we have an incomplete state net, but enough detail to begin to see more specifically some of the required behavior.

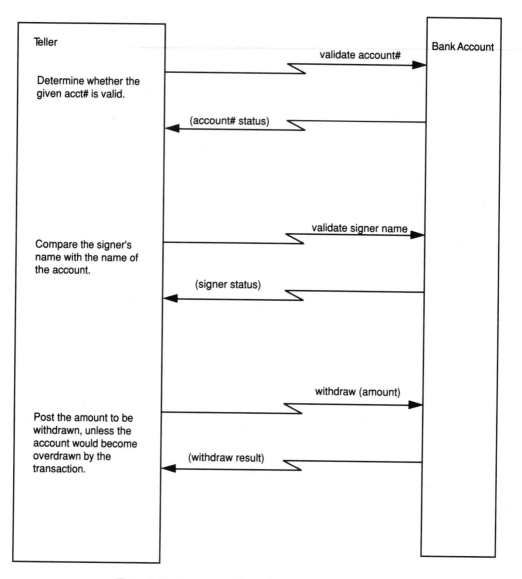

Figure 6.39   A sequence of interactions with behavior sketch of teller.

6.15.2 Interaction Descriptions for High-Level Interactions

When we create high-level interactions, we occasionally have to view interaction descriptions in a slightly different way. We usually consider an interaction description as an explanation of what takes place when an origin object interacts with a destination object. When we subsume several interactions together as a single interaction, however, some activity descriptions may describe interactions involving hidden objects and may, therefore, not make sense with the standard interpretation.

Figure 6.41 shows an example. In Fig. 6.41(a) a clerk prepares an unfilled order. Instead of depositing it in an out-basket as in our earlier example, the clerk gives the

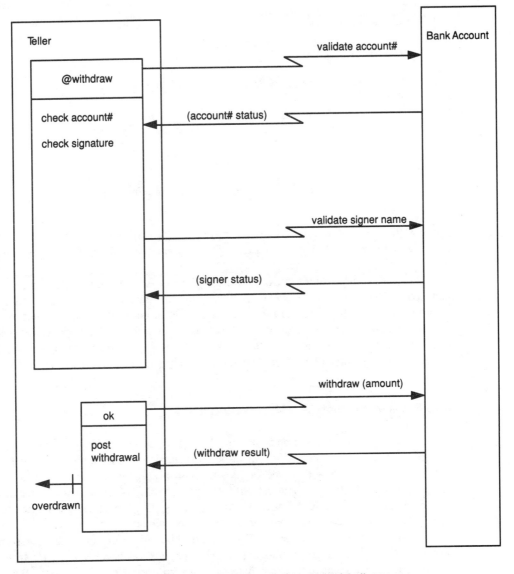

Figure 6.40 A sequence of interactions with partial teller state net.

unfilled order to an inventory clerk. The inventory clerk takes the order and prepares a pick list, which is a list of items to be taken from the bins to fill the order. The inventory clerk then gives the pick list to a packager.

Figure 6.41(b) shows how we subsume the intermediate inventory clerk under a high-level interaction link. The question here is what should the high-level interaction description be? When creating high-level interaction links from low-level interaction links, we use the interaction description of the destination object. Thus, in Fig. 6.41(b), the interaction description is *(pick list)*. If the interaction description defines the interaction between the origin and destination objects as in Figs. 6.1 and 6.2, we need do nothing more. Otherwise, we mark the interaction description with a dashed underline as

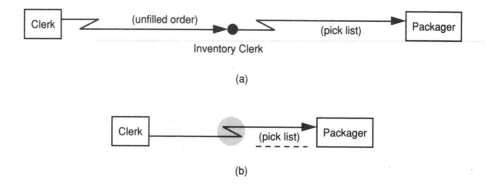

Figure 6.41   Dashed high-level interaction descriptions.

Fig. 6.41(b) shows. Here, a clerk does not give a *(pick list)* to a packager. Indeed, the clerk has nothing directly to do with pick lists—they are generated by an inventory clerk hidden in the high-level interaction.

Our dashed-underlining convention is similar to dashed underlines for constraints and notes. Recall that when we wish to display constraints and notes in high-level diagrams, but the information to evaluate a constraint or understand a note is not in the diagram, we mark the constraint or note by a dashed underline. Since we cannot properly interpret the interaction description in Fig. 6.41(b) without investigating the details of the high-level interaction link, we mark it with a dashed underline in a similar fashion.

6.15.3 Views for High-Level Interaction Links

Like high-level object classes and relationship sets for object-relationship models and high-level states and transitions for object-behavior models, we have high-level interaction links for object-interaction models. We have already shown some examples in Figs. 6.1, 6.11, 6.12, 6.19(b), 6.20, and 6.41(b). As with other high-level constructs, we have a shading convention to indicate that interaction links are high level, and we require that their interpretation be consistent with their atomic counterparts.

Figure 6.42 summarizes some of the possibilities for interaction-link views and also shows a high-level *n*-way interaction which has not yet been introduced. In Fig. 6.42(a) we have a basic interaction diagram with no high-level interaction links. A seller sets a minimum selling price for a piece of real estate. The seller's realtor then passes the seller's offer along to a buying realtor who gives it to a potential buyer. The buyer may respond with a counter offer, giving the maximum amount the buyer is willing to pay. The buyer's realtor then passes the buyer's offer along to the selling realtor who gives it to the seller. After several iterations, the sale may be negotiated.

In Fig. 6.42(b) we show a parallel construction by combining pairs of interaction links between the same two object classes. In Fig. 6.42(c) we show a linear construction by combining a sequence of interaction links so that only their end points are left. We may presume that we create the interaction diagram in Fig. 6.42(c) from the diagram in Fig. 6.42(b) and thus also see that we can subsume high-level interaction links in the creation of even higher level interaction links.

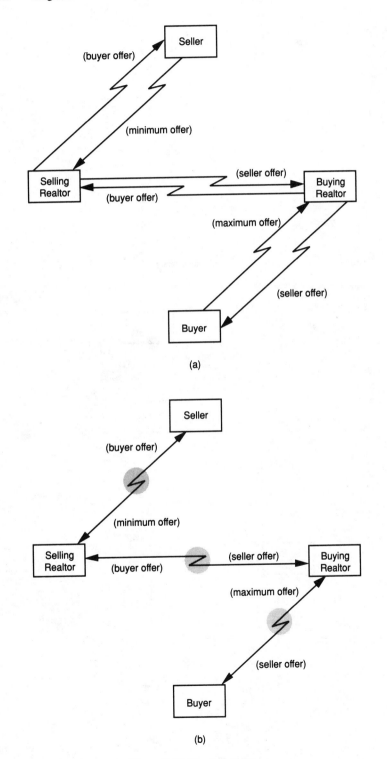

(a)

(b)

Figure 6.42 Interaction views.

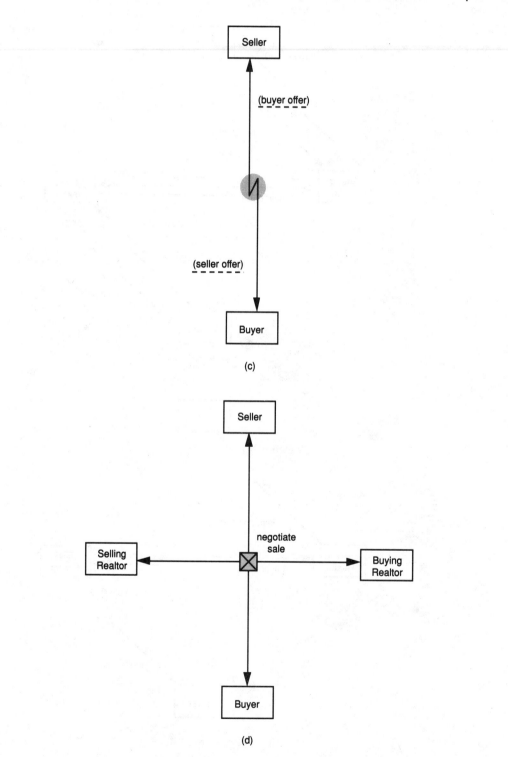

Figure 6.42 (Con't.)   Interaction views.

Figure 6.42(d) introduces a new diagram symbol for object interactions. Here, we begin with the original atomic interaction diagram in Fig. 6.42(a) and subsume all the interaction links, but none of the object classes. Since there are more than two object classes participating in the interactions, the result is a high-level *n*-way interaction. We denote high-level *n*-way interactions by overlaying two zigzags to form a box with diagonals. We then connect each of the *n* object classes participating in the interaction to the box as Fig. 6.42(d) shows. We shade the symbol to show that the interaction is high level.

The label on a high-level *n*-way interaction follows conventions we have already described. As we did in Fig. 6.20, we may place a label in the middle, near the interaction symbol, if it properly represents the activity of each of the *n* objects participating in the interaction. In Fig. 6.42(d) we have the interaction label *negotiate sale,* which can properly be viewed as the activity of each of the participants in the interaction.

6.16 Interactions and Generalization-Specialization

In earlier chapters, we discussed generalization and specialization as abstraction and organizational techniques for system modeling. Along with generalization and specialization, we also discussed inheritance. Inheritance for interactions applies as well. All objects of a specialization inherit all the interactions defined for objects in any of its generalizations. The objects of the specialization may have additional interactions defined; however, they must at least participate in all the interactions defined for objects of a more general class. Inheritance, of course, recursively applies up the generalization chain to the most general object class, or in the case of multiple inheritance, to the most general object classes.

Figure 6.43 shows an example. Here, *Bank Account* is a generalization of *Savings Account.* A bank customer interacts with all bank accounts by depositing and withdrawing money. Since a savings account is a special type of bank account, it inherits the interactions of a bank account. Hence, a bank customer can also deposit to and withdraw from a savings account. Notationally, these interactions are implicit and therefore need not be explicitly shown.

Objects in a specialization may also directly interact with other objects. Figure 6.43 shows that a bank customer may access a savings account to determine the current

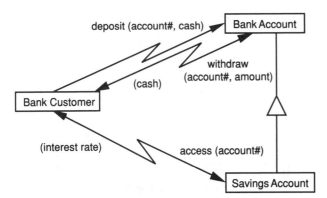

Figure 6.43   Interaction inheritance.

interest rate. Since not all bank accounts have an associated interest rate, this interaction cannot be done with *Bank Account*. Specialization object classes inherit interactions from generalization object classes, but not vice versa.

6.17 Sample Interaction Diagrams

We conclude our chapter on object interaction by giving two interaction diagrams that are more extensive and more complete than the examples given thus far. The first, in Fig. 6.44, shows an overall view of how the objects in the Green-Grow Seed Company interact. The second, in Fig. 6.45, shows the specifics of how packagers interact with other objects in the Green-Grow Seed Company. We explain these examples in this section.

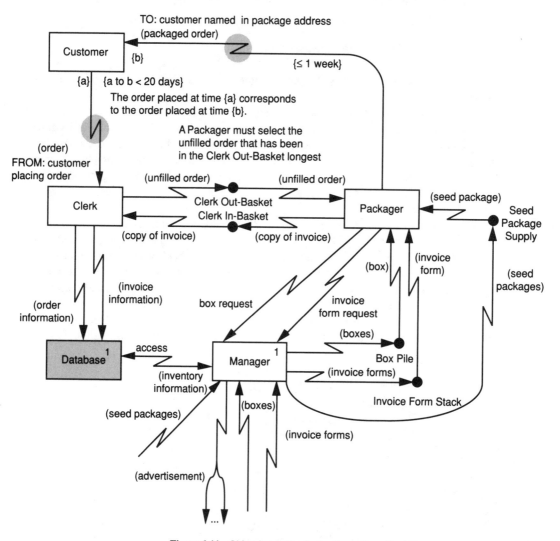

Figure 6.44   Object interaction in the Green-Grow Seed Company.

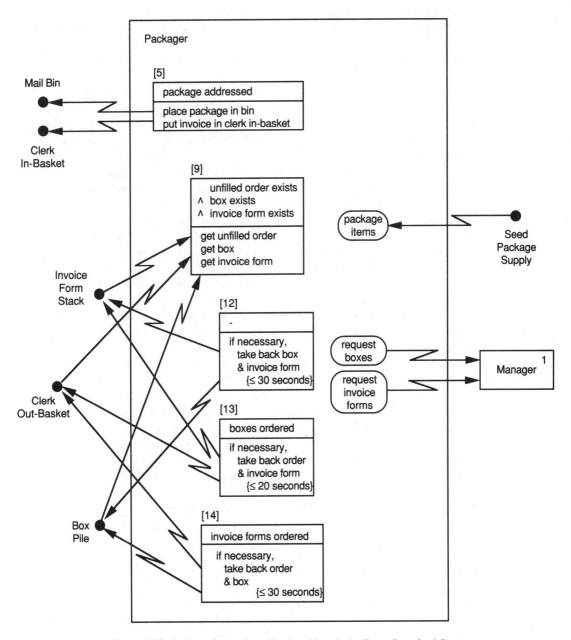

Figure 6.45   Packager interaction with other objects in the Green-Grow Seed Company.

6.17.1 High-Level Interaction Within the Green-Grow Seed Company

We begin in Fig. 6.44 with a customer placing an order. We see that an order goes from the customer to a clerk in the company and that the FROM-clause identifies the customer to whom the order should be shipped. The interaction link from *Customer* to *Clerk* is high level because it may involve a mail system, which is of no interest to our application and is, therefore, hidden within the link.

Upon receiving an order, a clerk creates order information for the database. The *Database* object class is shaded because it is high level. It includes many object classes and relationship sets that together describe the information stored in the company database. The clerk also places the unfilled order, which contains the order information, in the clerk out-basket.

When there are unfilled orders in the clerk out-basket, a packager gets one and fills the order. In taking an unfilled order from the clerk out-basket, a packager obeys the specified ordering constraint by taking the oldest unfilled order from the out-basket. To fill an order, a packager also needs a box, an invoice form, and seed packages. If there are no boxes or invoice forms, a packager can request them from the manager. When a packager finishes a packaging task for an order, the packager sends the packaged order to the customer named in the package address and puts a copy of the filled-in invoice form in the clerk in-basket. A clerk picks up the invoice copy and uses it to update the database by adding the new invoice information.

Since the packaged order goes through the mail, which we do not model, the interaction link from *Packager* to *Customer* is high level. Mailed packages should arrive back to the customer who placed the order within one week after being sent. The entire time from placement of an order ({*a*} in Fig. 6.44) to receipt of an order ({*b*} in Fig. 6.44) takes no more than 20 days.

The manager of the Green-Grow Seed Company makes sure that everything runs smoothly. The manager may access the database to check inventory information. When additional supplies of seed packages are needed, the manager obtains them and puts them in the seed package supply. The manager also gets boxes and invoice forms as needed. In addition, the manager is also responsible for advertising. Fig. 6.44 shows that the manager causes advertising to be broadcast to the outside world.

6.17.2 Packager Interaction within The Green-Grow Seed Company

Figure 6.45 shows the details of a packager's communication with other objects in the Green-Grow Seed Company. We see that in transition *9* a packager gets an unfilled order from the clerk out-basket, a box from the box pile, and an invoice form from the invoice-form stack. To fill an order, a packager gets seed packages from the seed package supply. In transition *5,* a packager places addressed packages in the mail bin and also puts filled-in invoice forms in the clerk in-basket.

If an exception occurs that makes it impossible to fill an order, then, if necessary, a packager places the unfilled order back in the clerk out-basket, the box back on the box pile, and the invoice form back on the invoice-form stack. When an exception occurs in transition *9* because there are no boxes, a packager sends a box request to the manager, and when an exception occurs because there are no invoice forms, the packager sends an invoice-form request to the manager. Figure 6.45 also illustrates, once again, how we can use partial state nets of object classes to describe object interaction.

6.18 Bibliographic Notes

Interaction models, whether process-interaction, data-interaction, or object-interaction models, are part of many analysis methods. Structured analysis, which uses a process-oriented interaction model, represents interaction by data flow diagrams [DeMarco 1979, Gane 1979, Hatley 1987, McMenamin 1984, Ward 1985, Yourdon 1989]. JSD, which is also a process-oriented interaction model, represents interaction by a system-specification diagram [Jackson 1983]. Other methods such as SRD use an interaction diagram to describe the context of the analysis [Orr 1981]. Both the process-oriented view and the data-oriented view of SADT are interaction models with constraints and mechanism or implementation added [Ross 1977]. OOA, which is object-oriented, uses an interaction model based on message connections that show the dependencies of one object on the services of another [Coad 1991]. OMT uses event traces to model event interactions among objects and uses process-oriented data flow diagrams to represent information interaction among objects [Rumbaugh 1991].

Allowing a rich and flexible description of interactions among objects instead of just among activities or processes is a primary difference between the OSA interaction model and most other interaction models.

6.19 Exercises

6.1 Figure 6.46 shows an interaction diagram. List all facts that are definitively known about the interaction depicted in the diagram.

6.2 Complete the object interaction diagram shown in Fig. 6.47 by giving all the interactions between a Customer and an ATM. Base the interactions on the behavior scenarios *ATM.1* through *ATM.9* given in the exercises of Chap. 3. Add individual states or transitions to the ATM only to show when the interactions should take place. It is not necessary to produce a fully connected ATM state net for this exercise.

6.3 The central computer of the KWE Bank has an Account Information Management System (AIMS) through which all account transactions requested at an ATM are processed. ATMs communicate with the AIMS through the ATM Network. Show the interactions between an ATM and the AIMS based on the behavior scenarios given at the beginning of Chap. 3 exercises. Provide partial state nets as needed.

6.4 All model KWE-2XL ATMs have a unique feature. Each unit calculates predictive maintenance metrics (PMMs) and can report these metrics on request from the central computer. If the metrics indicate that maintenance must be performed, the central computer then sends a message to the branch manager of the branch responsible for maintaining the ATM.

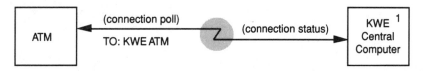

Figure 6.46   Interaction involving members of the KWE central computer and ATM object classes.

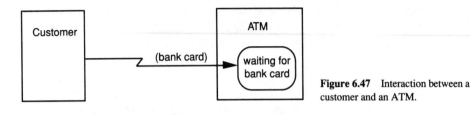

Figure 6.47 Interaction between a customer and an ATM.

The maintenance monitor software in the central computer sends a PMM pool request which is distributed to all KWE-2XL ATMs. All ATMs must respond by sending their PMM within 45 seconds. Each PMM response is kept in a response queue until processed by the maintenance monitor. Each PMM response queue entry has the PMM and the date and time at which the response was received. The central computer dequeues the responses for interpretation and possible notification of the branch manager.

Show the high-level interactions between the ATM, Maintenance Monitor, Central Computer, Response Queue, and Branch Manager without any state-net detail.

7

Model Integration

When an analysis task becomes too large for one individual, several analysts may work together to accomplish the task. Each analyst has responsibility for part of the system. The parts should be reasonably disjoint, and they should cover the entire system of interest. Once analysts have produced ORMs, state nets, and interaction diagrams for their assigned responsibilities, these components can be integrated to create a single ORM for the entire system with accompanying state nets and interaction diagrams as needed.

We present model integration by first giving an overall approach to integration for OSA models. Then, to illustrate model integration, we expand our running example. First we provide ORM diagrams for invoice information and inventory information and integrate them with the basic seed company ORM, which we developed in Chap. 2. We then provide state nets for clerks and for the manager and integrate them with the packager state net, which we developed in Chap. 3. Next, we discuss considerations for mixed integration of ORMs and state nets. Finally, we show how interaction diagrams lead to cross-component integration and system unification.

7.1 An Approach to Integration for OSA

The process of integration for OSA is similar to the process of schema integration in database systems. We directly adopt these database model integration techniques for ORMs. We also adapt these same techniques for integrating state nets and interaction diagrams. The order in which ORMs, state nets, and interaction diagrams are integrated does not matter. However, we usually integrate an ORM with another ORM first and then integrate state nets, class by class. Although we can also separately integrate interaction diagrams, we usually use interaction-diagram integration to put some finishing touches on ORM and state-net integration. The result is a complete and consistent integrated OSA model.

In addition to adopting and adapting database integration techniques, OSA allows us to make improvements on database integration. Using views, we may establish an integration framework and proceed top down. Top-down analysis encourages an appropriate divide-and-conquer approach to the analysis of large systems. Once a high-level view is in place, we are able to make reasonable decisions about how to divide the analysis task among a group of analysts so that each can develop a reasonably independent subpart for a large system. Top-down development also simplifies later integration and helps ensure that the integrated result will indeed model the entire system.

We begin with an integration framework, which guides the division of labor among the analysts as well as later integration of the diagrams produced by the analysts. Using the integration framework as a guide, we integrate diagrams in three steps. First, we compare diagrams. Second, we conform diagrams. Third, we merge diagrams. Before giving an extensive example, we make a few remarks about comparing, conforming, and merging diagrams.

7.1.1 Diagram Comparison

We compare diagrams to identify components that correspond. Corresponding components may or may not have conflicts. If there are conflicts, they will either be name conflicts or structural conflicts.

Name conflicts arise because of synonyms or homonyms. If the same component has two or more names in different diagrams, we have a synonym conflict. For example, we may have a state in one state net called *Filling Order* and the same state in another state net called *Packaging Seeds*. If different components have the same name, we have a homonym conflict. For example, *Date* may be the expiration date of a credit card in one ORM and the date that an order is shipped in another ORM.

Structural conflicts in the model arise either because different types of structures represent the same concept or because of conflicting constraints. Before listing common types of structural conflicts, we remark that attribute/object-class conflicts cannot arise in OSA. Since OSA does not require analysts to distinguish between attributes and object classes, the common difficulty of deciding whether something is an object or an attribute disappears. Therefore, it is impossible to have a component in one ORM described as an object class and in another ORM as an attribute.

One way type conflicts can occur is if a concept is described in one diagram at a high level of abstraction and in another diagram at a low level of abstraction. One analyst may, for example, have modeled the concept of customer discount information as Fig. 7.1(a) shows, while another may have modeled the concept as Fig. 7.1(b) shows. Here, the *qualifies for* relationship set in Fig. 7.1(b) captures the same idea as the relationship sets "*Preferred Customer Group has Name*," "*Customer is member of Preferred Customer Group*," and "*Preferred Customer Group has Discount Rate*" and the two object classes *Preferred Customer Group* and *Name* in Fig. 7.1(a). The diagram in Fig. 7.1(a), however, shows more detail because the analyst produced it at a lower level of abstraction.

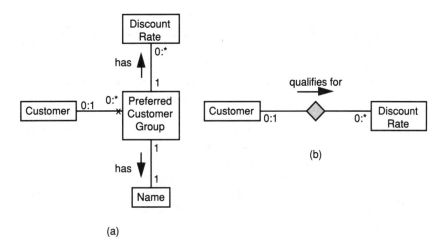

(a)

Figure 7.1   An abstraction conflict.

Type conflicts also occur when corresponding relationship sets have different structures. In one diagram, for example, a relationship set may be an *n*-ary relationship set while in another it is a part-of relationship set.

Constraint conflicts occur when there are inconsistencies in constraints. In Fig. 7.2(a), for example, object class *Order* in the "*Order sent on Date Shipped*" relationship set has a *1* participation constraint. This conflicts with the corresponding *0:1* participation constraint on *Order* in the "*Order sent on Date Shipped*" relationship set in Fig. 7.2(b).

As we have illustrated in Fig. 7.2, some constraints may be inconsistent because they are contradictory. Others may be inconsistent because they are incomplete. For example, one diagram may have a general constraint that is entirely missing in a corresponding diagram.

7.1.2 Diagram Conformance

Once we have identified conflicts, we resolve them. To resolve them, we conform diagrams. In the process of conforming diagrams, we must make sure that the resulting model remains consistent with reality. Sometimes, this means that we should go back to our sources of information to validate the newly conformed model.

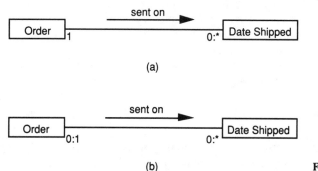

(a)

(b)   **Figure 7.2**   A constraint conflict.

We resolve synonym and homonym conflicts by changing names or allowing aliases as best suits our needs. We usually resolve homonym conflicts by selecting different names for all, or at least for all but one, of the homonyms in conflict. We may, however, also choose not to resolve homonym conflicts. Our OSA model does not assume, as many models do, that a reoccurrence of a name for the same construct makes the construct the same. For example, we may have *Name* of *Person* and *Name* of *Preferred Customer Group*. Here, *Name* and *Name* are homonyms and we may choose to leave them as they are. When we leave them the same, we have to disambiguate them by their context in the same way we disambiguate homonyms in everyday life.

Resolution of structural conflict is more difficult than resolution of name conflict. In general, diagrams must be made to conform by making alterations. We make alterations by adding, deleting, replacing, and combining diagram elements. The process is not automatic and usually requires insight and understanding by both analysts and clients.

7.1.3 Diagram Merge

Once diagrams conform, we can merge them. Two or more components considered to be conceptually identical become one component in the merged diagram. Diagrams obtained by merging should be complete and correct, but they are often not as minimal and not as understandable as we would like them to be. We may alter merged diagrams, however, so that they become reasonably minimal and more understandable.

We achieve minimality by removal of redundancy. There may, for example, be redundant relationship sets. In a merged diagram, for instance, we may have relationship sets that relate person to social security number, person to name, and name to social security number. The relationship set that relates name to social security number may be removed from the merged model, since it is implied by the other two. Useful redundancy need not be eliminated, and it is not necessary to achieve absolute minimality. In design, we are more interested in absolute minimality, but this need not be of great concern to us during analysis.

We achieve understandability by removal of unnecessary complexity and by restructuring diagrams into a more homogeneous global view. We may, for example, be able to generalize. If, for instance, on one diagram we have employees and on another diagram we have customers, we may be able to generalize them both as people in a merged diagram. A new object class, originating in neither diagram, may be added as a generalization and appropriate relationship sets may be established to provide information that may be inherited by the specializations. When treating people as a generalization of employees and customers, for example, we may attach components of information such as name, address, and telephone number to the object class for people and eliminate them from the object classes for employees and customers. We can recover them, of course, by inheritance, but the restructured diagrams will have become less complex and more expressive in the merge process.

7.1.4 Integration Policies

To complete model integration successfully for large systems, it is usually necessary to have integration policies. Integration policies may dictate which views take precedence

when conflicts arise and who has the final authority to resolve conflicts. The policies
may also prescribe the order of integration.

7.2 Integration Framework

To illustrate our approach to integration in OSA, we extend our seed-company example.
We assume that after some observation of the company in operation and after some discussion with the employees, we first obtain the high-level view of the company in Fig. 7.3.

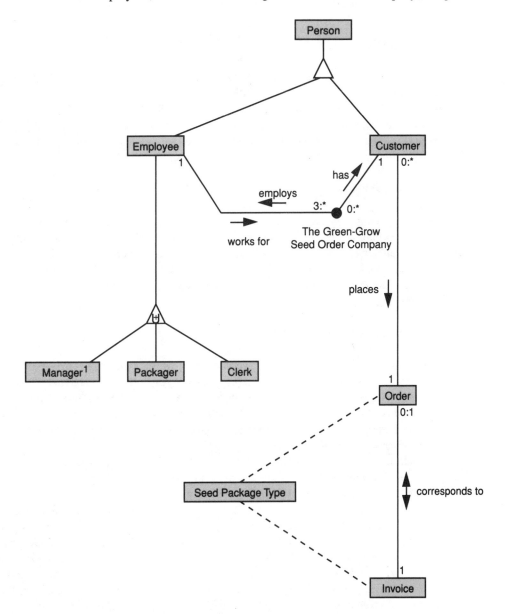

Figure 7.3   High-level integration conflict.

This high-level ORM serves as our integration framework. We see that a reasonable way to break up the analysis task is to assign analysts to develop ORMs for each of the high-level object classes in Fig. 7.3. We also realize that because of the generalization-specialization relationship among company personnel, we should probably do all the personnel together, perhaps using an intermediate high-level view if the task is complex. We also see that we probably need state nets for each of the different types of employees, for orders and customers, and perhaps also for invoices and seed packages.

7.3 ORM Integration

To illustrate ORM integration, we assume that an analyst interviews the manager about company personnel, obtaining the ORM in Fig. 7.4. Another analyst interviews a clerk about needed customer information, obtaining the ORM in Fig. 7.5. The analyst interviewing the manager also asks about seeds the company sells and obtains the ORM in Fig. 7.6. Two other analysts look respectively at the order forms and the invoice forms used by the company and obtain the ORMs in Figs. 7.7 and 7.8.

7.3.1 ORM Integration Strategy

After analyzing all these ORM views and considering the integration framework in Fig. 7.3, we conclude that the diagrams in Figs. 7.4 and 7.5 should integrate easily since there is almost no overlap and yet there is a significant interconnection because customers are people. We thus decide to integrate these two diagrams first. Afterwards, we will take the resulting integrated diagram and the remaining diagrams in Figs. 7.6, 7.7, and 7.8 and integrate them as one group.

7.3.2 Initial ORM Integration

In comparing Figs. 7.4 and 7.5, we identify several synonyms. *Name* of *Person*, *Address* of *Person*, and *Discount* in Fig. 7.4 are respectively synonyms of *Customer Name, Customer Address,* and *Discount Rate* in Fig. 7.5. We can also identify several homonyms. *Name* of *Person* in Fig. 7.4 and *Name* of *Preferred Customer Group* in Fig. 7.5 are homonyms, and *Date* on which an employee received a Bonus in Fig. 7.4 and *Date* of last order in Fig. 7.5 are homonyms.

In looking at the diagrams, we surmise that "*Employee gets Discount*" is a high-level relationship set that contains the discount information in Fig. 7.5. A quick check with either the manager or a clerk confirms our guess.

We conform the two diagrams as follows. To resolve the homonym conflicts, we decide to do nothing about the *Name*s, but decide to replace *Date* on which an employee received a bonus by *Bonus Date* and *Date* of last order by *Last-Order Date*. We do not need to resolve the synonym conflicts *Customer Name* and *Name* and *Customer Address* and *Address* because we will have no need for *Customer Name* and *Customer Address* when we merge the diagrams and make *Customer* a specialization of *Person*. We thus remove *Customer Name* and *Customer Address* in preparation for

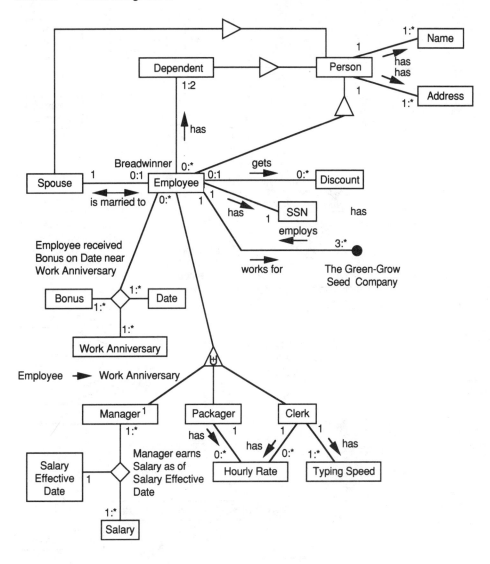

Figure 7.4 Personnel view.

the merge. We also remove *Discount* in Fig. 7.4 since the information is contained in the discount information in Fig. 7.5.

To merge these altered diagrams, we only need to attach *Customer* as a specialization of *Person*, as the integration-framework diagram in Fig. 7.3 shows. The merged diagram of 7.4 and 7.5, which is complete, correct, and reasonably minimal is in Fig. 7.9.

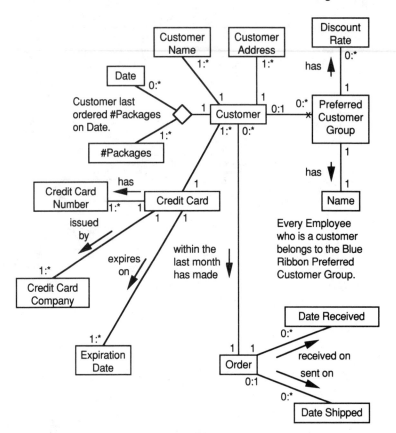

Figure 7.5   Customer view.

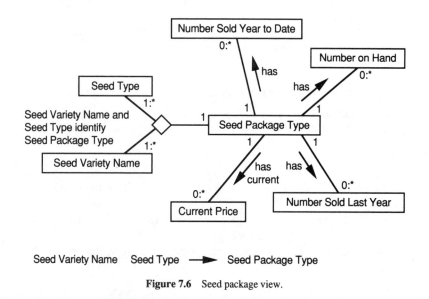

Figure 7.6   Seed package view.

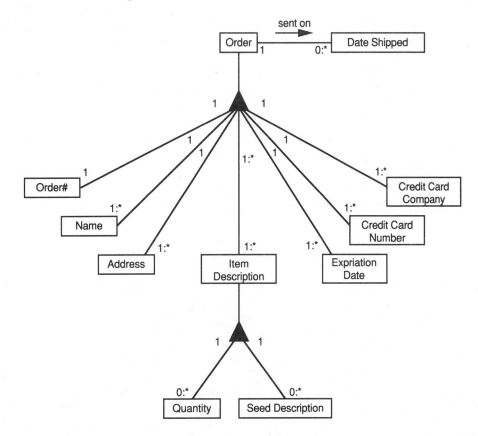

Figure 7.7   Order view.

7.3.3 Completion of ORM Integration

We now integrate the diagrams in Figs. 7.6, 7.7, 7.8, and 7.9. In comparing these diagrams we observe conflicts and resolve them as follows. *Quantity* in Fig. 7.7 is a synonym of *Quantity Requested* in Fig. 7.8, and *Current Price* in Fig. 7.6 is a synonym of *Unit Cost* in Fig. 7.8. To resolve these synonym conflicts, we replace *Quantity* by *Quantity Requested* and *Current Price* by *Unit Cost*.

There is a structural conflict since *Seed Type* and *Seed Variety Name* in Fig. 7.6 together correspond to *Seed Description* in Figs. 7.7 and 7.8. To resolve this structural conflict, we replace the object classes *Seed Type* and *Seed Variety Name* by *Seed Description* and replace the associated relationship set "*Seed Variety Name and Seed Type identify Seed Package Type*" by "*Seed Description identifies Seed Package Type.*"

The participation constraints for *Order* in "*Order sent on Date Shipped*" are in conflict in Figs. 7.7 and 7.9 because the minimum for *Order* in Fig. 7.7 is *1,* but for Fig. 7.9 is *0.* To resolve this participation-constraint conflict, we replace the participation constraint for *Order* in Fig. 7.7 with *0:1.*

We are now ready to merge the four diagrams in Figs. 7.6, 7.7, 7.8, and 7.9. Except for the object class *Name*, we consider all object classes and relationship sets having the same name to be conceptually identical. We thus combine the conceptually identical object classes *Quantity Requested, Seed Description, Unit Cost, Date Shipped,*

Order, Address, Discount Rate, Credit Card Company, Credit Card Number, and *Expiration Date.* Of the object classes called *Name,* we consider those that participate in the relationships sets *"Name is subpart of Order," "Name is subpart of Invoice,"* and *"Person has Name"* to be conceptually identical, and therefore, combine them.

The only relationship sets in the four diagrams that have the same name are called *"Order sent on Date Shipped."* We thus combine *"Order sent on Date Shipped"* in Fig. 7.5 with *"Order sent on Date Shipped"* in Fig. 7.9.

Having merged these diagrams, we recognize some simplifications we can make. We realize that there is a one-to-one correspondence between *Invoice* and *Order,* and every invoice has a corresponding order, but not every order has a corresponding invoice. We also realize that if we know the order (or the invoice), then we know the customer, and we thus know the *Name, Address, Credit Card Company, Credit Card*

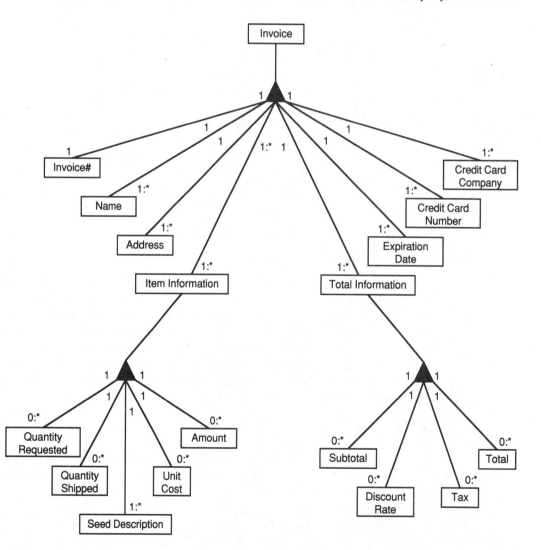

Figure 7.8   Invoice view.

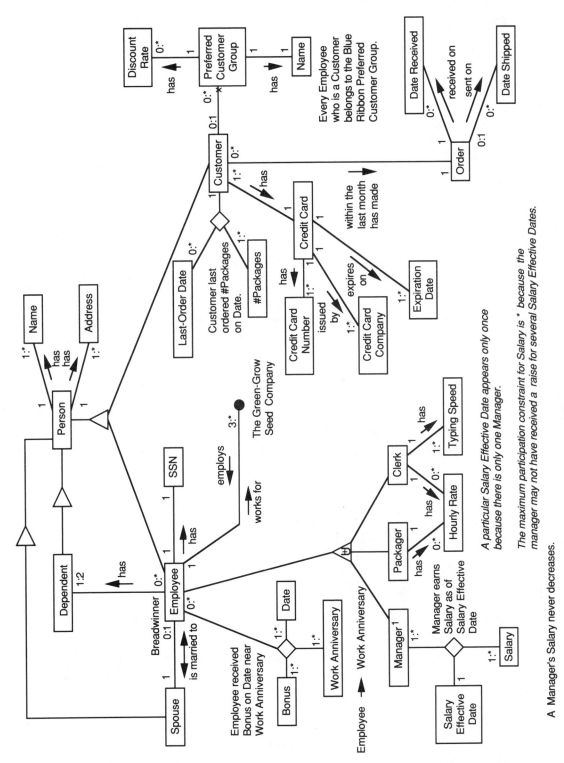

Figure 7.9 Integrated employee-customer view.

215

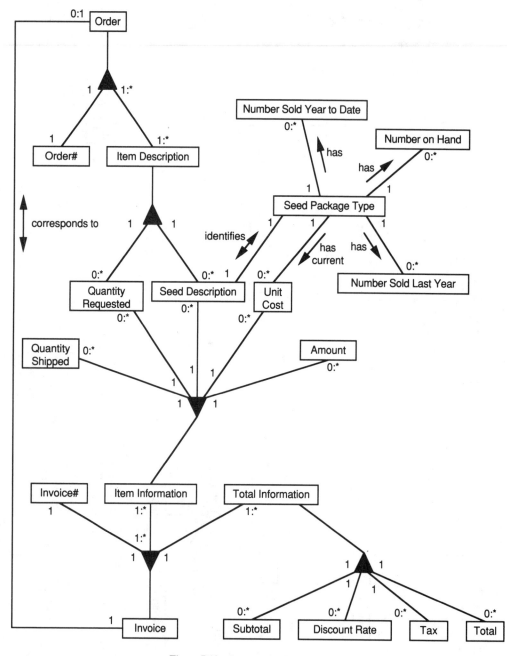

Figure 7.10 Integrated order-invoice-seed view.

Number, *Expiration Date*, and *Discount Rate*. We thus add the relationship set "*Order corresponds to Invoice*" with participation constraints *0:1* for *Order* and *1* for *Invoice*, and we remove relationship sets connecting both *Order* and *Invoice* to *Name*, *Address*, *Credit Card Company*, *Credit Card Number*, *Expiration Date*, and *Discount Rate*.

These simplifications lead to a single integrated ORM represented by the two diagrams in Figs. 7.9 and 7.10. The only common component in the integrated ORM in Figs. 7.9 and 7.10 is the object class *Order*. The two diagrams are viewed as a single ORM diagram with the two representations for the *Order* object class superimposed.

7.3.4 High-Level ORM Integration Views

Although complete, correct, and reasonably minimal, the resulting diagram from the previous example is unwieldy. This is why we have shown it in two parts in different diagrams, rather than as one diagram. When analysts work with large ORM diagrams, they are not likely to view them all at once. Instead, they are likely to use various view diagrams that show only a small part of a large diagram. Within the small diagram, components in the analyst's focus of attention are likely to be low-level components. Related contextual components are likely to be displayed at a higher level, and other components are likely to be omitted.

Since the ORM in Figs. 7.9 and 7.10 is the complete ORM for the seed-order company, there should be a correspondence between it and the high-level integration-framework diagram in Fig. 7.3. Indeed, we can create the high-level view in Fig. 7.3 from the ORMs in Figs. 7.9 and 7.10 as follows. We create a dominant object class for *Person* that includes *Name* and *Address*. We create a dominant object class for *Employee* from Fig. 7.4 by cutting through the *is a* relationship sets and by excluding *Discount* since it was discarded in the integration process and excluding *The Green-Grow Seed Company* since we wish to have it in our result. We also create from Fig. 7.4 dominant high-level object classes for *Manager, Packager,* and *Clerk* by including in each their regular, attached relationship sets. We create a dominant object class for *Seed Package Type* by including all the object classes associated with it in Fig. 7.10. We create a dominant object class for *Order* and for *Invoice* by including the part-of hierarchy for each of them as given in Fig. 7.10. Finally, we create a dominant object class for *Customer* by including the object classes in Fig. 7.5 except *Order, Date Shipped,* and *Date Received.*

If we implode all these high-level object classes, we would essentially have the diagram in Fig. 7.3. To make the imploded diagram exactly the same as Fig. 7.3, we would have to discard a dashed line between *Packager* and *Clerk* that arises because both packagers and clerks are connected to the *Hourly Rate* object class. Also, we discard a dashed line between *Employee* and *Person* that arises because of the spouse and dependent connections. Notice that we have a different name for the relationship set between *Customer* and *Order,* which we would have to rename. If we make these changes, we have Fig. 7.3. In addition, we have established the connection between the low-level diagrams in Figs. 7.9 and 7.10 and the high-level view in Fig. 7.3.

7.4 State-Net Integration

To illustrate state-net integration, we assume that three analysts are working together to develop a state net for a clerk's behavior. The main responsibility for the first analyst is to learn how a clerk processes a letter. We assume that this analyst produces the state net shown in Fig. 7.11. The second analyst is responsible to learn about updating information from a completed invoice form and produces the state net shown in Fig. 7.12. The third analyst is responsible to learn how a clerk handles a phone call and produces the state net shown in Fig. 7.13.

7.4.1 State-Net Integration Strategy

After looking at these three views of clerk behavior, we see that the diagrams in Figs. 7.11 and 7.12 have considerable overlap and are likely to merge easily. We decide to integrate the diagram in Fig. 7.11 with the diagram in Fig. 7.12 first and then to integrate the result with the diagram in Fig. 7.13.

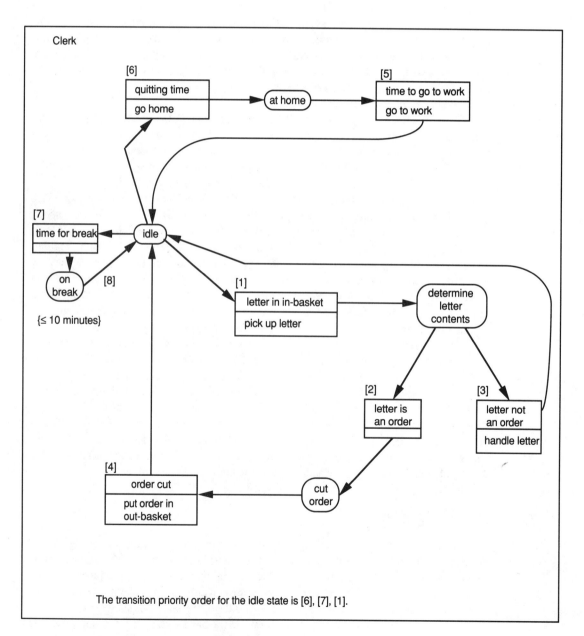

Figure 7.11   State net for processing a letter.

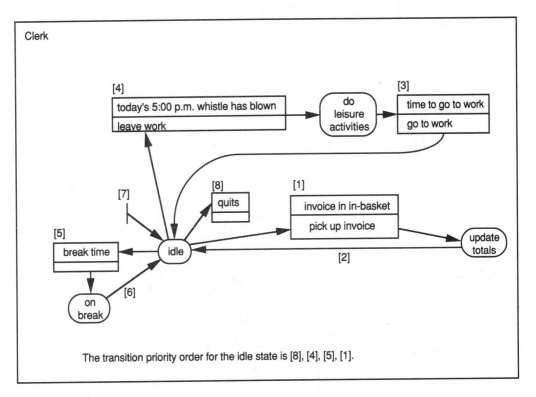

Figure 7.12   State net for updating invoice information.

7.4.2 Initial State-Net Integration

In comparing the diagrams in Figs. 7.11 and 7.12, we conclude that the *idle* states in both are the same state. We also conclude that transition *5* in Fig. 7.11 and transition *3* in Fig. 7.12, which have identical triggers and actions, are the same. Several other states and transitions are also the same, having different but synonymous labels, triggers, and actions. In addition, we observe that in Fig. 7.12 a clerk initially starts in the *idle* state and may quit. These initialization and termination transitions have been omitted (probably overlooked) by the analyst who developed the state net in Fig. 7.11. On the other hand, we observe that in Fig. 7.11 there is a ten minute time limit for a break. This has been omitted (probably overlooked) by the analyst who developed the state net in Fig. 7.12. We also observe that the priority constraints need adjusting.

To make the diagrams conform, we decide to adopt the notation in Fig. 7.12. Thus, in Fig. 7.11 we rename states *at home* and *break,* respectively, to be *do leisure activities* and *on break*. We also revise the notation of transitions *6* and *7* in Fig. 7.11 to be identical, respectively, to transitions *4* and *5* in Fig. 7.12. Finally, we add the object creation and destruction transitions that appear in Fig. 7.12, but were omitted in Fig. 7.11, and add the real-time constraint of Fig. 7.11, but omitted from Fig. 7.12. To make the priority constraints conform, we ask the clerks, who tell us that processing a letter has priority over processing an invoice. Since these priority constraints will only make sense in the merged diagram, we remove them in preparation to merge the diagrams.

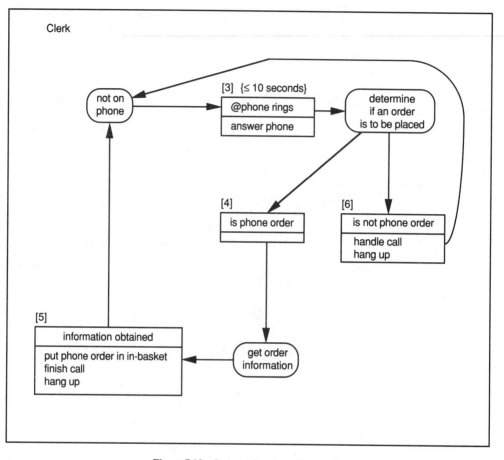

Figure 7.13   State net for processing a phone call.

Since we have now made the diagrams conform, we can merge them. In merging diagrams, transition identifiers are ignored, but we can arbitrarily choose new ones and add them to the diagram. Using these newly chosen transition identifiers, we can add the priority constraint we need. Fig. 7.14 shows the result.

7.4.3 Completion of State-Net Integration

Now we complete our state-net integration by considering the state nets in Figs. 7.13 and 7.14. When we compare the state net in Fig. 7.13 with the state net in Fig. 7.14, we encounter some interesting and subtle questions. The analyst working on phone calls understands that whenever a clerk is not on the phone and the phone rings, the clerk answers the phone, but is this always true? Consideration of the state net in Fig. 7.14 shows that it is not, since the clerk, for example, may be home doing leisure activities. The question of whether a clerk on break answers a phone also arises and is resolved by asking a clerk, who may respond "No, we only answer a phone when we are working, but not on the phone."

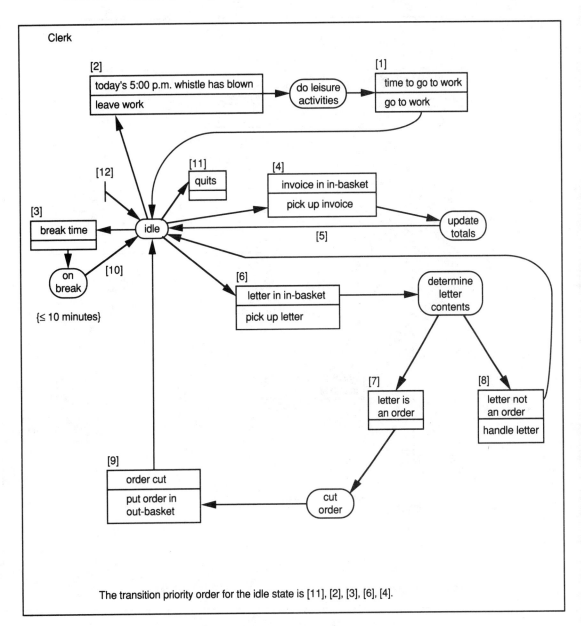

Figure 7.14   State net for processing a letter and updating invoice information.

Another interesting and subtle question that arises is what a clerk does after a phone call. It might make sense to return to the *cut order* state shown in Fig. 7.14 and cut the order received on the phone. Although this seems reasonable, there are problems with this solution. When a clerk answers the phone, the clerk may have been in any one of several states. Furthermore, the phone may ring again while a clerk is cutting an order for the previous phone call. To find out how a clerk resumes processing after a phone call, we consult with the clerks and learn that after information for a phone order

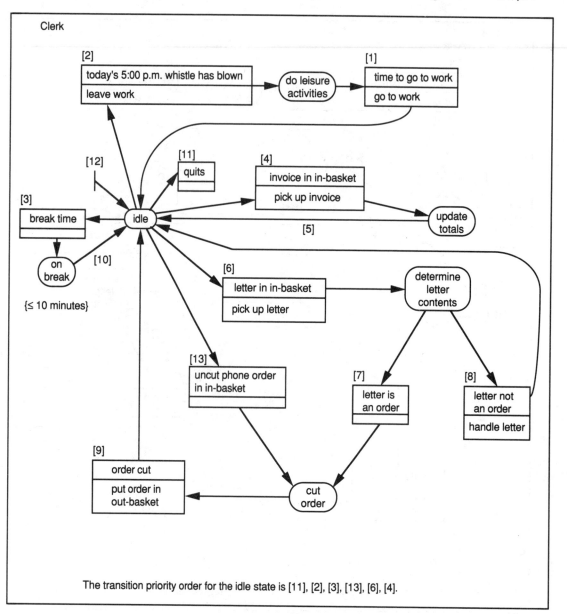

Figure 7.15   State net for not on phone.

is obtained, a clerk puts the information in the in-basket and resumes working in the state that was interrupted by the phone call.

Having discovered these conflicts and answered questions that have surfaced, we are now able to conform the diagrams. We first add a new transition in Fig. 7.14, that fires when a clerk is idle and an uncut phone order is in the in-basket. The priority order for this transition is higher than the priority for processing a letter, but lower than the priority for taking a break. In the diagram in Fig. 7.15, transition *13* is the new transition added to the diagram in Fig. 7.14.

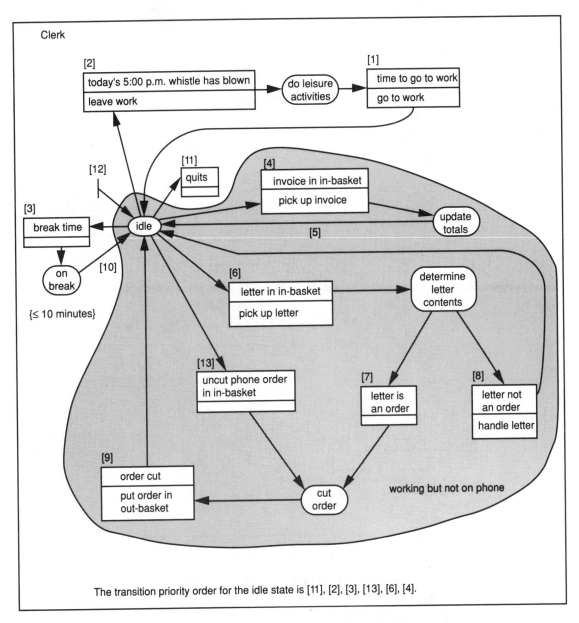

Figure 7.16   Identification of substate net for working but not on phone.

To allow a phone call to interrupt a clerk only if the clerk is at work and not on break, we define the high-level state *working but not on phone* shown in the cloud in Fig. 7.16. To make the diagrams in Figs. 7.13 and 7.16 conform, we rename the high-level state *not on phone* in Fig. 7.13 to also be *working but not on phone*. We now implode the high-level state in Fig. 7.16 and merge the resulting diagram with the modified diagram in Fig. 7.13. Fig. 7.17 shows the result.

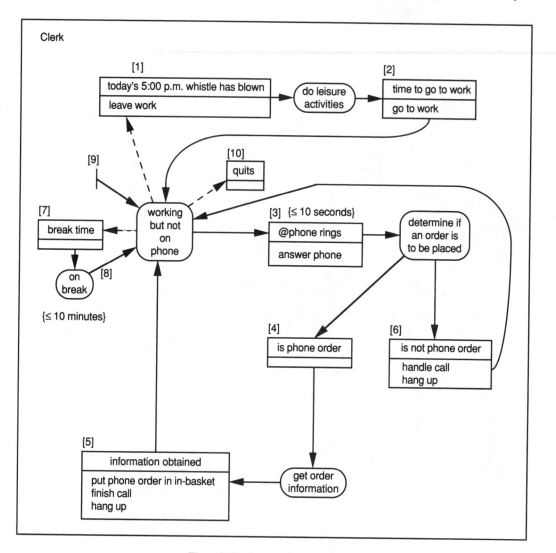

Figure 7.17   Integrated state net for clerk behavior.

The new high-level state *working but not on phone* defined in Fig. 7.16 may be given separately as the low-level state net in Fig. 7.18. In Fig. 7.18 we have added the constraint

A Clerk resumes at the interrupted state

to override the entry specification and cause a clerk to resume in the state interrupted by a phone call.

7.4.4 State-Net Generalization

Having finished state nets for clerks and having earlier completed a state net for packagers, we may be interested in a state net that describes the general behavior of an

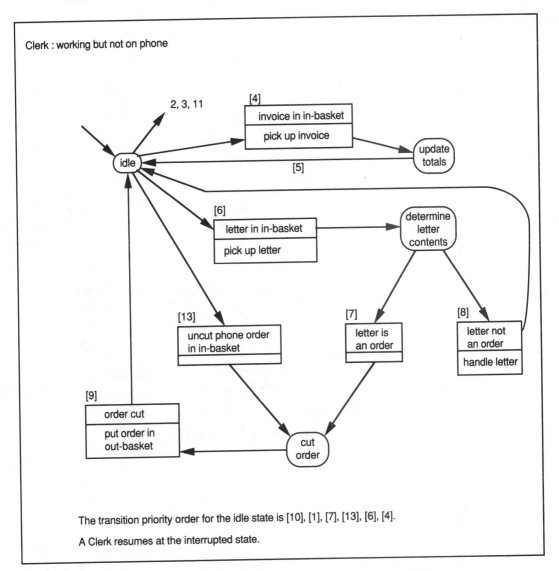

Clerk : working but not on phone

2, 3, 11

[4]
invoice in in-basket
pick up invoice

update totals

idle

[5]

[6]
letter in in-basket
pick up letter

determine letter contents

[13]
uncut phone order in in-basket

[7]
letter is an order

[8]
letter not an order
handle letter

[9]
order cut
put order in out-basket

cut order

The transition priority order for the idle state is [10], [1], [7], [13], [6], [4].

A Clerk resumes at the interrupted state.

Figure 7.18 Integrated state net for working but not on phone.

employee. If we had a state net for the manager, we would be able to extract the common behavior of all employees to form the state net. We, therefore, assign an analyst to create a state net for the manager.

Figure 7.19 shows the result. The state net in Fig. 7.19 shows that a manager at work is in one of six states: *idle, working on inventory, working on advertising, getting seed packages, getting boxes,* or *getting invoice forms.* Transitions among these states take place when the manager decides to change from one activity to another. As do other employees, the manager leaves work to do leisure activities and returns when it is time to go to work.

Figure 7.20 shows the state-net generalization for an employee. We see that all employees are hired, and they quit (presumably when fired, retired, or otherwise). They

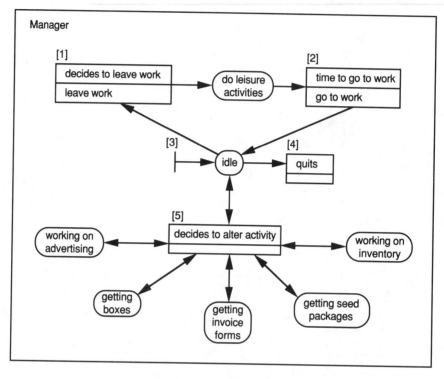

Figure 7.19   Manager state net.

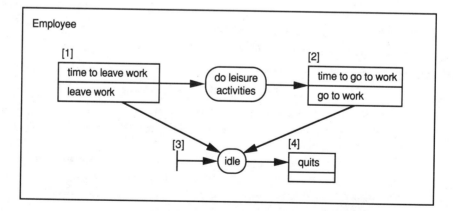

Figure 7.20   Employee state net.

also go home when it is time to leave work to do leisure activities and return to work at the appropriate time. To make the generalization state net properly correspond to the specialization state nets, we have chosen a new trigger in transition *1* of the employee state net. The trigger condition *time to leave work* for an employee leaving work is appropriate since it is weaker than the trigger condition *decides to leave work* for a manager and weaker than the trigger condition *today's 5:00 p.m. whistle has blown* for a clerk and a packager. As discussed in Chap. 3, a trigger for a generalized transition as in the employee example must be either equivalent to or weaker than corresponding triggers in state nets of specialized object classes.

7.5 Mixed ORM and State-Net Integration

It is possible for one analyst to represent a system component in an ORM and another analyst to represent the same component in a state net. It is also possible for a component or configuration of components in an ORM to suggest the need for a state net, and vice versa, for a component or configuration of components in a state net to suggest the need for additional ORM components. We must, therefore, consider both ORMs and state nets together as part of OSA integration.

For example, we have no state net for *Order,* and yet the ORM description for an order suggests that an order can be in one of several states. In Fig. 7.9 shown previously, we see that there can be a relationship instance between an order and a date shipped in the *"Order sent on Date Shipped"* relationship set. We thus know that an order can be in the state of having been shipped. Similarly, the relationship set *"Order received on Date Received"* implies that an order can be in a state of having been received. Between the states *received* and *shipped,* the order is in a state of being processed, which is likely to be a high-level state with several substates. If we consider the state nets for a clerk and a packager, we can learn about the details of these additional states for an order.

Based on these observations, we may decide to develop a state net for *Order.* Figure 7.21 shows the result. An order comes into existence when it arrives. The order is then in the state of having been received. When a clerk selects a received order to prepare an order form, the order changes from the *received* state to the *being cut* state. After the order is cut, it waits to be filled. While it waits to be filled, it is also in the *in Clerk Out-Basket* state. When a packager selects an order to be filled, the order enters the *being filled and packaged* state. On rare occasions, there might not be boxes or invoice forms. When this happens, the order must go back into the out-basket and wait again to be filled. Once the order is filled and packaged, the order is shipped. The order continues to exist as an object in our model of the Green-Grow Seed Company until one month after it is shipped.

The analyst is responsible for making ORMs and state nets complete and consistent. On examining the *Order* state net in Fig. 7.21, for example, we see a reference to boxes, but boxes are not part of our ORM for the Green-Grow Seed Company. To make our model consistent and complete would require that we add a *Box* object class. Any needed relationship sets would also be added.

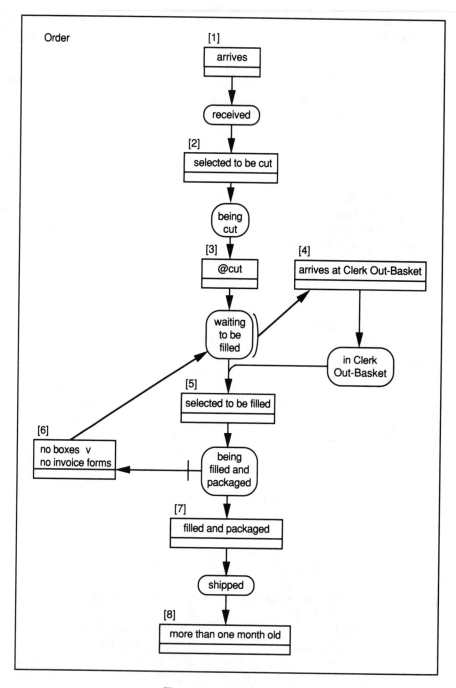

Figure 7.21   Order state net.

7.6 Integration Considerations for Object Interaction

Interaction diagrams also play a role in model integration. We may, for example, assign an analyst to determine the high-level interactions within the Green-Grow Seed Company. The result might be the interaction diagram in Fig. 7.22, which we developed in Chap. 6. From a high-level interaction diagram we may discover new object classes, the need for additional state nets, and the need to adjust existing state nets.

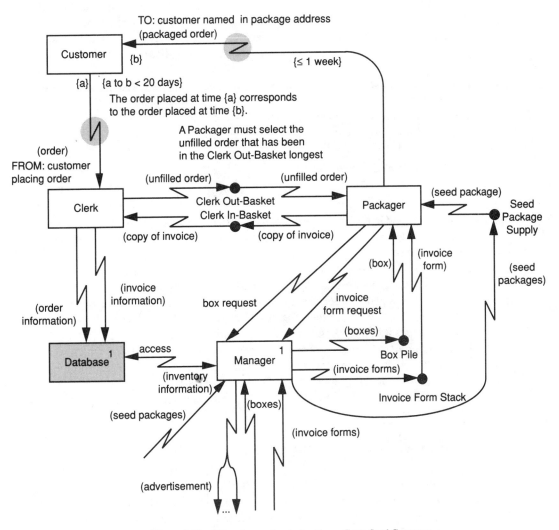

Figure 7.22   Object interaction in the Green-Grow Seed Company.

7.6.1 New Object Classes

Interaction diagrams may help us discover new object classes. In Fig. 7.22, for example, there are several object classes that do not appear in the ORM in Figs. 7.9 and 7.10. Most of them represent intermediate repositories for object interaction. These include the singleton object classes *Seed Package Supply, Box Pile, Invoice Form Stack, Clerk Out-Basket,* and *Clerk In-Basket.* We may add these intermediate repositories to an ORM as stand-alone object classes. If we wish, we may also establish any appropriate relationship sets for them. For example, we may show ownership of the clerk in- and out-baskets by establishing an owns relationship set between the repositories and the *Clerk* object class.

The only object class in Fig. 7.22 that is not a repository and does not appear in the ORM in Figs. 7.9 and 7.10 is the object class *Database.* This raises the question of what information is stored in the company database. With a little reflection and through discussions with company personnel, the analysts would come to understand that all the information in the ORM in Fig. 7.9 and 7.10 should constitute the database. We therefore create an independent high-level object class that encompasses the entire ORM and name it the *Database* object class. Since there is only one database, it is a singleton object class. Since it is a singleton object class, every object and relationship instance in the high-level object class belongs to it and there is thus no need to create any relationship sets connecting it to its low-level object classes.

7.6.2 Additional State Nets

Figure 7.22 shows, among other things, the interaction between a customer and internal elements of the Green-Grow Seed Company. However, we do not have a state net for a customer. If we wish to show how a customer interacts with components of the Green-Grow Seed Company, we need to supply a customer state net.

Figure 7.23 shows a customer state net and also shows a customer's interaction with the Green-Grow Seed Company. A customer either places a phone order or a written order. For a phone order, the order message flows to a clerk from the *making phone order* state. For a written order, the customer writes out the order in the *writing order* state and sends it to a clerk in the *mail order* action of transition 3. The interaction link for a mail order is high-level because it passes through a mail system, which we are not concerned with in our application. A customer receives a packaged order in the *pick up packaged order* action in transition 1. The interaction link for a packaged order is also high level, because it also passes through the mail system, the details of which are of no interest for our model.

7.6.3 Adjustments to Existing State Nets

When we integrate interaction diagrams with existing state nets, we check consistency by looking for a state or transition for every interaction arrow. We can check this type of consistency by creating an interaction diagram involving a state net as we did for *Customer* in Fig. 7.23. Instead of creating the diagram from scratch, however, we use the existing state net and make adjustments as necessary.

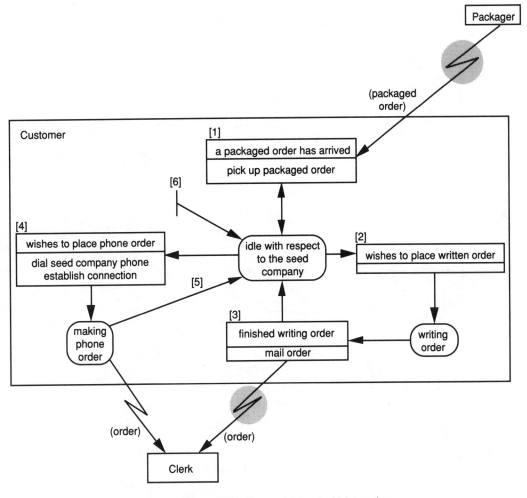

Figure 7.23   Customer state net with interactions.

Figure 7.24 shows an example using the state net of the manager as given in Fig. 7.19 and the high-level interaction diagram in Fig. 7.22. In the interaction diagram in Fig. 7.24 we immediately have a transition or state for all the interaction arrows except the *box request* and the *invoice form request* going from a packager to a manager. With a little reflection, we realize that these requests should reach the manager whenever the manager is in any state or transition while at work. We thus create the high-level state *at work* shown in the cloud in Fig. 7.24. We are then able to direct these request interaction links from a packager to the manager's *at work* state, as Fig. 7.24 shows.

We can make the same kind of consistency check for object class *Clerk* by creating an interaction diagram for the clerk state net in Figs. 7.17 and 7.18. Figure 7.25 shows the result. Here no adjustments are necessary. Figure 7.25(a) shows the interaction link for a phone order from a customer. Figure 7.25(b) shows the interaction link for a written order from a customer and also shows the interaction links to the database and the clerk's in- and out-basket.

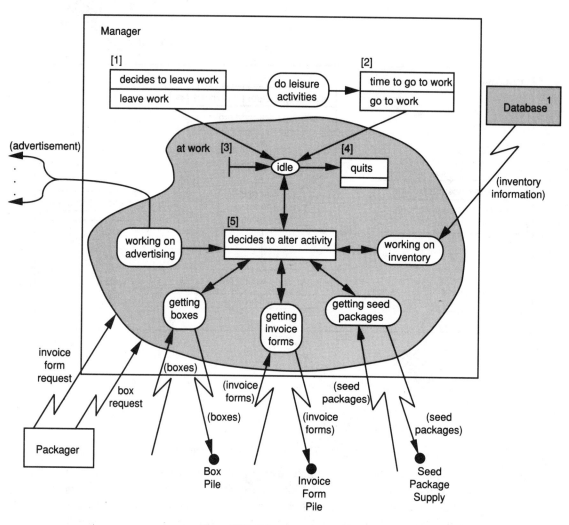

Figure 7.24   Manager state net with interactions.

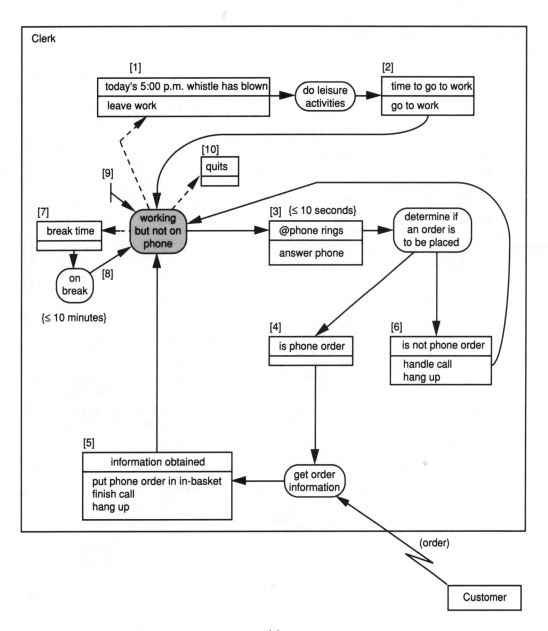

(a)

Figure 7.25   Integrated clerk state net with interactions.

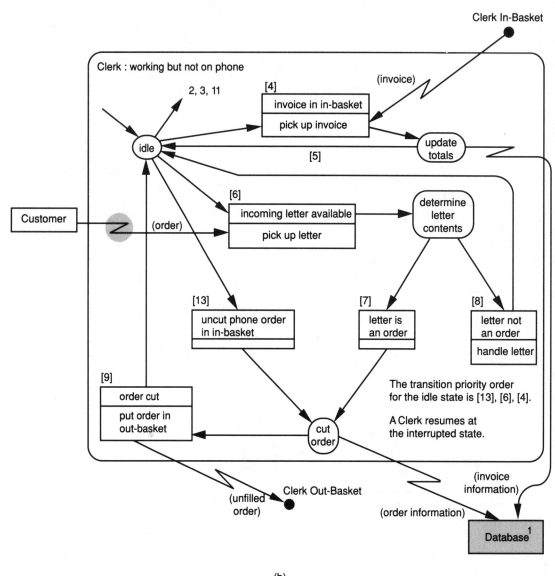

Clerk In-Basket

Clerk : working but not on phone

2, 3, 11

(invoice)

[4]
invoice in in-basket
pick up invoice

idle

update totals

[5]

[6]
incoming letter available
pick up letter

determine letter contents

Customer

(order)

[13]
uncut phone order in in-basket

[7]
letter is an order

[8]
letter not an order
handle letter

The transition priority order for the idle state is [13], [6], [4].

A Clerk resumes at the interrupted state.

[9]
order cut
put order in out-basket

cut order

Clerk Out-Basket

(unfilled order)

(order information)

(invoice information)

Database[1]

(b)

Figure 7.25 (Cont'd.)   Integrated clerk state net with interactions.

7.7 Summary Remark

Model integration is an important activity. It lets us merge independently developed models into a single, accurate view of a system. Not only is the process useful for developing a single view, but the process also requires analysts to resolve many issues regarding model completeness and accuracy prior to using the model for system design.

7.8 Bibliographic Notes

For a good survey of database schema integration and a comparison of several different methods, see [Batini 1986]. Several database books also discuss schema integration. For example, see [Elmasri 1989].

7.9 Exercises

7.1 Produce ORM diagrams for *The KWE Bank, Bank Personnel Customer, Signature Card, Account, Bank Card, Automated Teller Machine (ATM),* and *New Year's Club* as described in Chap. 1. Integrate all these ORM diagrams into a unified, understandable, and reasonably minimal ORM. (As a result of previous exercises, many of these ORM diagrams will have already been produced. Some of them may have also been integrated. Make use of these results in this exercise.)

7.2 Produce a state net for the 14 scenarios described in Chap. 3 and integrate it with the state net for an ATM inquiry transaction in Fig. 4.20. (If Exercise 3.1 has been completed, state nets for the 14 scenarios will have already been produced. Use them in this exercise.)

7.3 Develop interaction diagrams for the integrated ORM diagrams and state nets from Exercises 7.1 and 7.2. (Use any interaction diagrams developed for the exercises in Chap. 6 as an aid to completing this exercise.) If necessary, revise and extend the ORM diagrams and state nets developed for Exercises 7.1 and 7.2 so that the final OSA model is complete and properly documents the KWE banking system as presented.

8

OSA Model — An Example

In this chapter we use OSA to model a real-time system rather than a data processing system, which we commonly used for illustration in earlier chapters of the book. The system we analyze here is an amateur radio satellite communication system. We do not present an exhaustive description of the system, but present enough to show the use of OSA in a real-time example.

8.1 Amateur Radio Satellite Communication System

An amateur radio operator is an individual licensed by a country to communicate with other amateur radio operators on specific radio frequency bands. One way amateur radio operators communicate is by voice and data communication through a satellite.

Figure 8.1 illustrates how amateur radio operators communicate through satellites. When a transmitting operator wishes to say something to a receiving operator, the transmitting operator speaks into a microphone attached to a radio. The radio generates and sends a signal to the antenna, which then transmits the signal to a satellite visible to both operators. The satellite rebroadcasts the signal which is picked up by the receiving-operator's antenna, sent to the radio, and then sent to a set of headphones worn by the receiving operator.

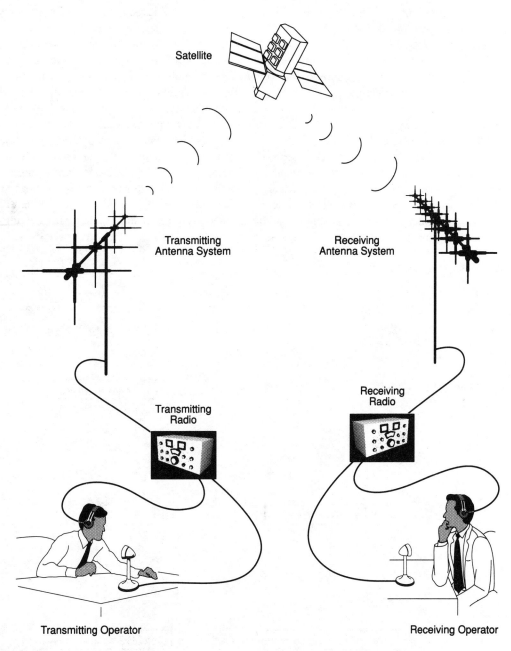

Satellite

Transmitting
Antenna System

Receiving
Antenna System

Receiving
Radio

Transmitting
Radio

Transmitting Operator

Receiving Operator

Figure 8.1   Amateur radio satellite communication.

A difficult problem for an amateur radio operator communicating through a satellite is keeping the antennas aimed at the satellite. An amateur-radio satellite continuously moves relative to the surface of the earth, and the satellite's movement varies according its orbital position. The best quality communication is achieved when antennas are aimed directly at a satellite. Amateur radio operators who first experimented with satellite communication aimed antennas manually. Today, many operators use computerized systems to aim antennas automatically. The example we present in this chapter documents a portion of a computerized satellite communication system.

The object-interaction diagram in Fig. 8.2 shows the basic objects in our model and also shows a high-level view of the interaction among the objects. The basic object classes are *Amateur Radio Operator, Radio, Antenna System, Satellite, Satellite Tracking System,* and *Satellite Information System.* Notice that an operator directly interacts with only two parts of the system, the radio and the satellite information system. The satellite information system, rather than the operator, is responsible for interacting with the satellite tracking system, which directs the antenna system to aim its antennas at a satellite. Notice also that the interaction between the satellite information system and the satellite tracking system is one-directional. The information system provides information for the satellite tracking system, but not vice versa.

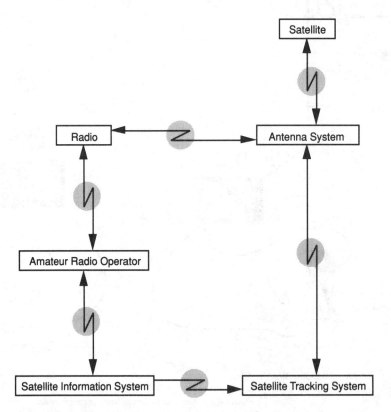

Figure 8.2   Object-interaction diagram for satellite communication.

The following sections describe some of objects that appear in the system model in more detail. As usual, we use ORMs to present descriptive information about the system, state nets to describe behavior, and object-interaction diagrams to show how objects within the system interact.

8.2 Satellite Information System

The ORM in Fig. 8.3 shows the *Satellite Information System* as a singleton independent high-level object class. The information system can run on a *Computer* that has a *Display Screen* and a *Real-Time Clock*. The real-time clock is assumed to be running synchronously with real-world time. The information system has a *System Clock* that runs at a specified *Rate* and has the current *System Time*. The system clock does not necessarily run synchronously with the real-time clock. A system clock's time may be

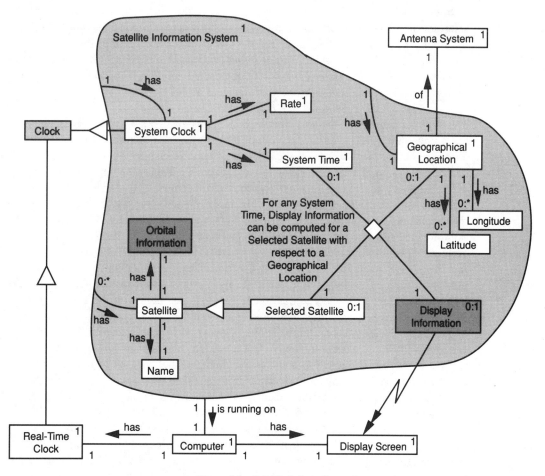

Figure 8.3   Satellite information system.

different, and it may run faster or slower than a real-time clock. The information system also has a *Geographical Location* that gives the *Latitude* and *Longitude* of an *Antenna System*. The object class *Satellite* in the information system contains zero or more satellites of interest, one of which may be chosen as the *Selected Satellite*. For each satellite, the information system knows its *Name* and its *Orbital Information*. Given a selected satellite, the orbital information for the selected satellite, and the geographical location of the antenna, the information system can compute *Display Information* describing the satellite's position with respect to the antenna's location for the current system time. The constraints ensure that the display information exists when there is a selected satellite.

The interaction link between the information system and the display screen in Fig. 8.3 indicates that display information continuously flows to the computer's display screen. The display information is a function of time, which changes continuously. Hence, the display information on the screen continuously changes.

Display Information in Fig. 8.3 is a high-level object class. Figure 8.4 shows its details. There are two parts to the display information: *Graphical Information* and *Textual Information*. The ORM in Fig. 8.4 shows the *Graphical Information* for the selected satellite as an independent high-level object class that includes a *Map*, a *Satellite Footprint*, a *Sub-Satellite Point*, and the *Geographical Location* of the antenna system. The ORM also shows the *Textual Information* as an independent high-level object class that includes the *System Time, Name of Selected Satellite, Time to Next Rise, Time to Next Set, Mode, Azimuth, Elevation, Range,* and *Altitude*.

In Fig. 8.4 the constraints $a + b = 1$, $a = 1$ when *Elevation* < 0, and $b = 1$ when *Elevation* ≥ 0 along with the participation constraints a and b tell us that a satellite either has a time until it next rises or a time until it next sets, but not both. If a satellite is not above the horizon with respect to the antenna system, we display the time to its next rise. On the other hand, if a satellite is above the horizon, we display the time to its next set.

Many of the display-information terms are technical. To clarify them, an OSA analyst may use notes and constraints. Figure 8.5 shows some ORM notes that clarify the graphical-display information. In particular, we see that a sub-satellite point is the point on the earth directly under a satellite and that a footprint is the area on the earth's surface that is visible from a satellite. Figure 8.6(a) shows a note that clarifies the meaning of altitude, elevation, and range with respect to the geographical location of an antenna system and a subsatellite point. It also defines a triangle whose sides are the altitude, the range, and the chord distance between two points on the earth's surface. Figure 8.6(b) gives a constraint that must hold since the law of cosines holds for all triangles. The equation in Fig. 8.6(b) is one of many constraints that would be needed to describe a satellite system completely. Other needed constraints would include formulas for the chord-distance function and the chord-angle function used in the constraint in Fig. 8.6(b) as well as numerous equations from classical orbital mechanics. Figure 8.7 shows some notes that clarify the meaning of azimuth, which is an angle from true North in a clockwise direction. Figure 8.8 explains what a mode is by listing some of the modes in a table and by stating that a mode is a shorthand name used to designate which receiving and broadcasting frequencies a satellite uses. In addition to clarifying these technical terms, Figs. 8.5 to 8.8 also show that we can use drawings, illustrations, maps, and tables as notes in OSA and that we can use mathematical formulas and equations as constraints.

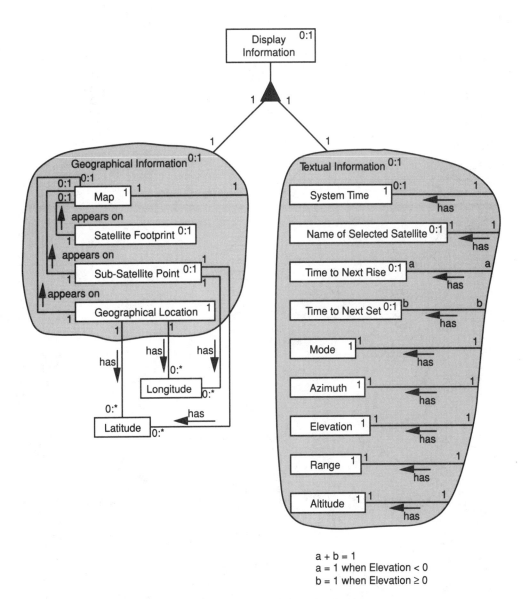

a + b = 1
a = 1 when Elevation < 0
b = 1 when Elevation ≥ 0

Figure 8.4   Display information details.

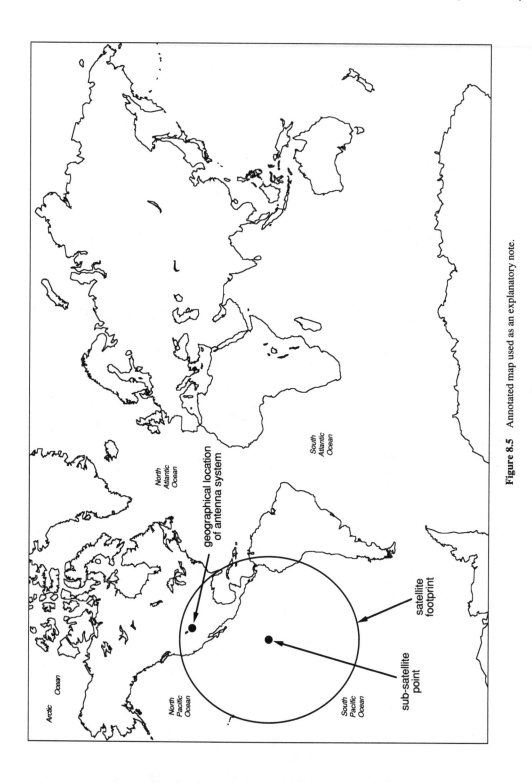

Figure 8.5 Annotated map used as an explanatory note.

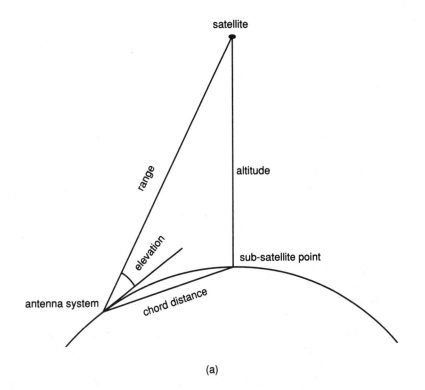

(a)

$$range^2 = altitude^2 + (chord\_distance(antenna, sub\text{-}satellite\_point))^2 -$$

$$2*altitude*chord\_distance(antenna, sub\text{-}satellite\_point)*cos(angle(sub\text{-}satellite\_point))$$

(b)

Figure 8.6   Related notes and constraints.

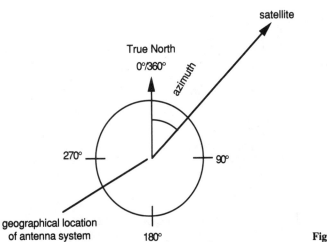

Figure 8.7   Azimuth description.

Mode	Satellite Receiving Frequency (MHz)	Satellite Broadcasting Frequency (MHz)
A	146	29
B	435	146
J	146	435

A mode designates a satellite's receiving and broadcasting frequency.

Figure 8.8   Mode table.

8.3 Real-Time and Simulated-Time Clocks

The object class *Clock* in Fig. 8.3 is a dominant high-level object class. Figure 8.9 shows its details. All clocks have a *Time, Rate,* and *Granularity* associated with them. The time of a clock is the current time of the clock. The current time changes at every granularity instance according to a nonnegative rate multiplier. When the rate multiplier is zero, the clock is stopped.

A *Real-Time Clock* is a specialization of *Clock.* A real-time clock has the current local time, properly offset from Greenwich Time, and runs synchronously with Greenwich Time, with a rate of one. Its granularity is as coarse or fine as is needed.

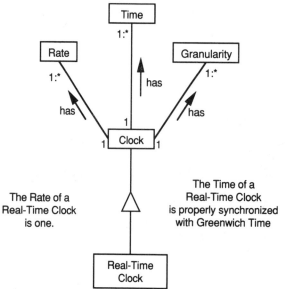

The Rate of a Real-Time Clock is one.

The Time of a Real-Time Clock is properly synchronized with Greenwich Time

Figure 8.9   Clock details.

As illustrated in Fig. 8.4, we can establish an ORM for both simulation and real-time modeling. In real-time modeling we use a real-time clock. For simulation we can change the time, rate, and granularity of a clock to achieve our purposes. For instance, the satellite information system displays both graphical and textual information that is changing with time. By synchronizing the system clock with local time, we can view the changes in satellite information as it occurs. On the other hand, we can set the time of the system clock forward or backward to see things that will happen or have happened. We can also slow down, speed up, or stop the clock to observe changes in satellite information at these different rates.

8.4 Satellite Tracking System

In addition to the task of displaying current satellite information, the satellite information system is also responsible for providing data to the satellite tracking system. An amateur radio operator uses the information system to find a satellite for communication. Once found, the operator asks the satellite information system to initialize tracking. Upon receiving a request to initialize tracking, the information system sends a tracking table to the satellite tracking system. The diagram in Fig. 8.10 shows these interactions.

Figure 8.11 shows an ORM for a *Satellite Tracking System*. A satellite tracking system is a singleton independent high-level object class that has access to a *Real-Time Clock* and is connected to a *Rotor System*. Within the high-level object class, we see that a satellite tracking system has at most one *Tracking Table*. A tracking table is a list of *Tracking Element*s where each element contains a *Time,* an *Azimuth,* and an *Elevation.* The list is ordered by time, and the difference between successive time values is Δ Time.

Figure 8.12 shows a state net for the tracking system. Initially, the system is in the *standby* state. While in the *standby* state, it may be turned on. When the satellite system is turned on, its behavior depends on whether the satellite to be tracked is visible. If the satellite is not visible, the system enters the *on* state. While in the *on* state, if the time of a tracking element is equal to the time of the real-time clock, then, transition *3* occurs. During transition *3*, the tracking system issues a *move to azimuth(X)* command and a *move to elevation(Y)* command where X and Y are respectively the azimuth and elevation of the tracking element whose time is equal to the time of the real-time clock. These commands control the *Rotor System* that aims the antennas at the satellite. If the satellite system is turned on while the satellite is visible, the antennas are to be initialized by aiming them at the azimuth and elevation of the most recent element in the tracking table. Commands to do the initializing are sent to the *Rotor System*. After initialization, the system enters the *on* state. If the system is turned off while in the *on* state, the system re-enters the *standby* state. If the tracking system is in either the *on* or the *standby* state and a tracking table arrives, then, as transition *4* shows, the tracking table of the tracking system changes.

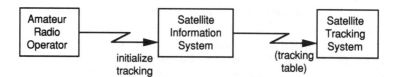

Figure 8.10   Initialization of the satellite tracking system.

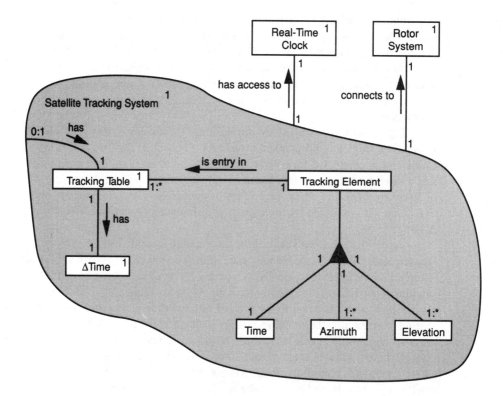

The Tracking Table is ordered by Time.

Successive Times differ by ΔTime.

Figure 8.11   Satellite tracking system.

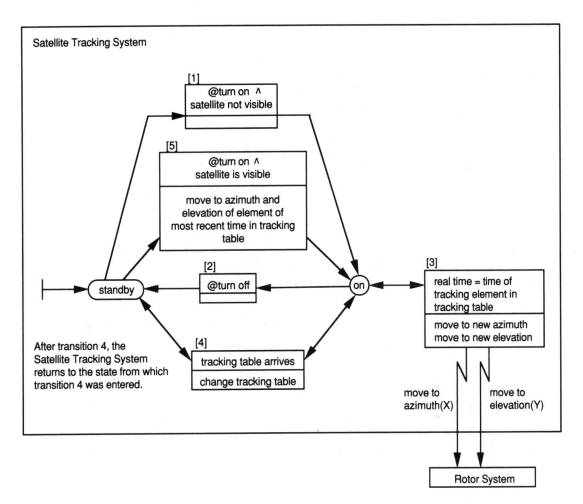

Figure 8.12   State net for the satellite tracking system.

8.5 Antenna System

Figure 8.13 shows the ORM for the *Antenna System* as a singleton independent high-level object class. The antenna system has a *Boom* that holds from one to four *Antenna*s. Each antenna can transmit or receive on one *Frequency Band*. A frequency band is a set of frequencies between some lower and some upper frequency inclusive.

The boom of the antenna system is clamped to a *Rotor System* which is mounted to a *Stand* or *Tower*. The rotor system is composed of a *Controller* and two rotors, an *Azimuth Rotor* and an *Elevation Rotor*. The azimuth rotor consists of an *Azimuth Brake*, a *Left Switch*, and a *Right Switch*. The elevation rotor consists of an *Elevation Brake*, an *Up Switch*, and a *Down Switch*. The ORM shows that an antenna system has an *Azimuth* and *Elevation* which are the azimuth and elevation of the respective rotors. The ORM

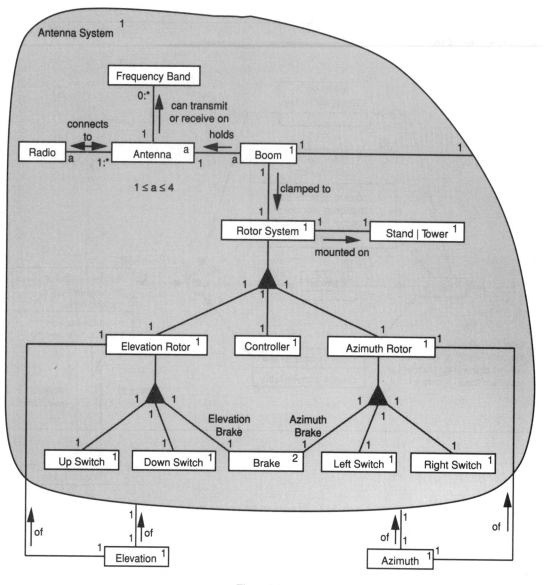

Figure 8.13 Antenna system.

also shows that an antenna connects to one or more *Radio*s. A radio connects to each of the antennas on the boom, between one and four depending on how many the boom holds.

We have found that the behavior of a high-level object often mirrors the behavior of one of its components. This is the case with the antenna system whose behavior is determined by the behavior of the rotor system. Furthermore, the rotor system is an aggregate whose behavior can be understood by understanding the composite behavior of its three major subcomponents: the controller, the elevation rotor, and the azimuth rotor.

Figure 8.14 shows a state net for the controller's basic behavior. Initially, the controller is in the *off* state. When the controller is turned on, it enters both the *ready to move in azimuth* and *ready to move in elevation* states. When the controller is turned off, it re-enters the *off* state.

When in the *ready to move in azimuth* state, the controller responds to the command *move to azimuth(X)*. Similarly, when in the *ready to move in elevation* state, the controller responds to a *move to elevation(Y)* command. The object interaction links included in the state net show that the *move to azimuth(X)* and *move to elevation(Y)* commands come from the satellite tracking system.

The state net in Fig. 8.14 also shows that we have chosen to model the process of moving to a new elevation or azimuth as actions in transitions instead of actions in states. We chose the transition-action model because we expect movement to an elevation or azimuth to be an atomic operation that should always complete without interruption. If an interruption occurs, we expect it to occur only as the result of an exception condition.

As the real-time constraints in the diagram show, we expect transitions *3* and *4* to take no longer than 30 seconds to complete. If either transition takes longer than 30 seconds, an exception occurs. When an exception occurs in transition *3* or *4*, the requested movement should be suspended and the controller must shut down the rotors to protect the antenna system against damage. To model this, we show appropriate exception conditions on each state and transition controlling the rotors in normal operation. These exception conditions ensure that if one rotor is blocked (i.e., does not complete a requested movement within 30 seconds), both rotors are turned off. Transition *5* ensures that all rotor switches are turned off if an exception condition occurs.

An operator normally does not turn the controller off while the antennas are moving, but it can happen. When it does, we consider it to be an exception. For this reason, we have included the shutdown condition on the exception arrows for transitions *3* and *4*. All the exception conditions together guarantee that rotor switches turn off gracefully whenever an exception occurs.

We now investigate the details of action behavior for transitions *3* and *4*. Since high-level transitions *3* and *4* are similar, we only describe the details of transition *3*. Figure 8.15 shows the state net providing these details. Once the controller receives an azimuth-move command, the controller enters the *release azimuth brake* state, which is the initial state in the substate net for transition *3*. While in this state, the controller sends a *release* command to the azimuth brake. When the brake releases, the controller enters the *ready to move* state. Based on the new position *X,* the controller then either turns on the left switch or the right switch and enters the *azimuth rotor moving left* state or the *azimuth rotor moving right* state. The controller stays in one of these states until the antenna system reaches the new azimuth position. When the new position is reached, the left or right rotor switch is turned off and the controller enters the state *wait for azimuth rotor spin down*. When the motor has spun down, the controller enters the *applying azimuth brake* state. In this state the controller sends an *apply* command to the azimuth brake. Finally, the controller exits the substate net and returns to the *ready to move in azimuth* state in Fig. 8.14.

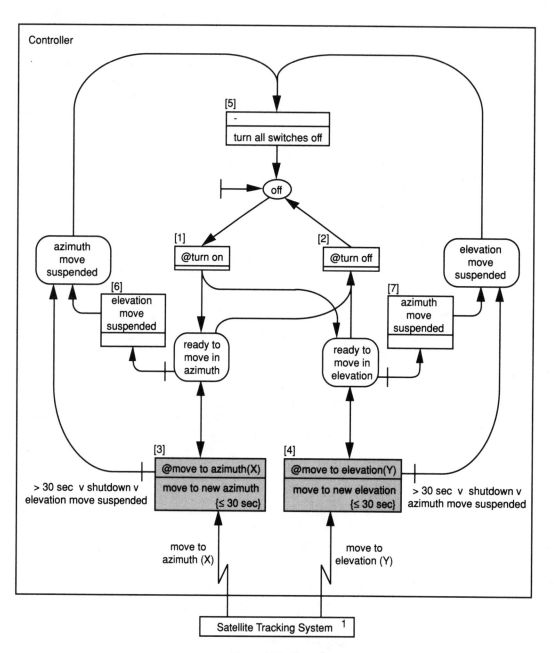

Figure 8.14   Controller state net.

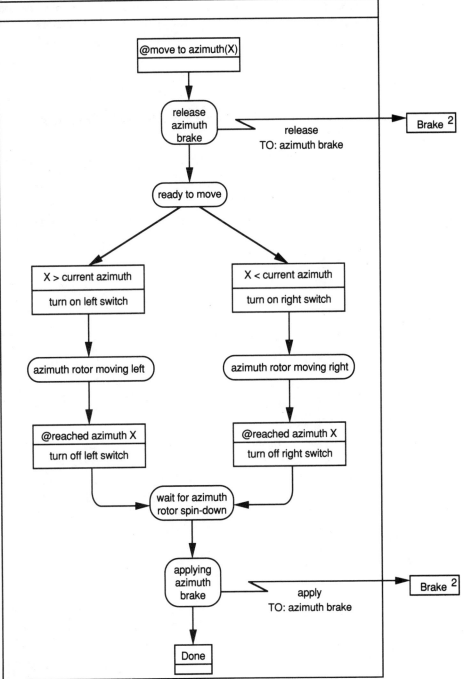

Figure 8.15   State net for a controller moving to a new azimuth.

8.6 Summary Remark

We used OSA in this chapter to model a portion of a real-time system with a reasonable amount of complexity. This example, coupled with earlier examples, demonstrate that the modeling components of OSA may be used to capture the semantics of a broad range of systems.

8.7 Bibliographic Notes

For those interested in more information about the satellite tracking system used in this chapter, see [Davidoff 1990].

Appendix A

A Formal Definition of OSA

This appendix formally defines OSA. In particular, it formally defines what it means for a model instance to be a valid OSA model instance.

As part of the formal definition, we give an OSA meta-model in terms of ORM diagrams. The OSA meta-model should be enlightening to all readers, even those without the background necessary to understand all the formalism. We thus present it first, even though it should come last in the formal presentation. Figures A.1 through A.4 comprise the ORM for ORMs; Figs. A.5 through A.10 comprise the ORM for state nets; and Figs. A.11 through A.14 comprise the ORM for interaction models. Each of these figure groups constitutes a single ORM constructed by superimposing object classes on the different diagrams that have the same name. Together these diagrams informally define the set of all valid OSA model instances.

In the remainder of this appendix we show that, with a proper mathematical interpretation, these diagrams also formally define the set of all valid OSA model instances. From here on, this appendix assumes reader familiarity with first-order predicate calculus and mathematical model theory.

In our approach to formally defining valid OSA model instances, we begin by providing a mapping from an Object-Relationship-Model instance to a first-order language and a set of first-order rules. In our formalism, language predicates will be well-formed formulas whose variables are all free, and rules will be well-formed formulas whose variables are all bound. We then map the first-order language to a mathematical model instance consisting of a universe of objects, a set of relations, a set of constants, and a set of functions. Using this latter mapping, we can check the validity of the mathematical model instance by ascertaining the truth of the rules. If all the rules hold, the ORM instance is valid.

To formally define OSA model instances, we then use the meta-model for OSA in Figs. A.1 through A.14. From the meta-model, we derive predicates and rules that must hold for any OSA model instance. All valid mathematical model instances for the OSA meta-model are valid OSA model instances. Hence, the OSA meta-model defines the set of valid model instances for OSA.

As an aid to understanding, we illustrate our definitions with some examples. Figure A.15 shows a slightly-altered version of a small part of our seed-company ORM; Fig. A.16 shows a small abstract ORM diagram created specifically to explain several points of interest; and Fig. A.17 shows an aggregation hierarchy. We use these examples to illustrate how we map an OSA model instance to a first-order language and a set of first-order rules. We also use the abstract ORM as a sample instance for our ORM meta-model.

A.1 Mapping from an ORM to First-Order Predicates and Rules

Our first task is to map an OSA model instance S to a first-order language and a set of first-order rules. Before beginning, however, we must make some slight adjustments in some cases. We allow homonyms for objects, object classes, relationships, and relationship sets in ORMs, but for our formal definition, we want these names to be unique. Therefore, before beginning we alter, where necessary, the names of objects, object classes, relationships, and relationship sets so that they are all distinguishable. We may for example, add subscripts, 1 for the first homonym, 2 for the second, and so forth.

A.1.1 Predicates

Predicates are derived from objects, object classes, relationships, and relationship sets of S as follows.

1. Except for dominant high-level object classes, each object class of S maps to one one-place predicate for each of its names. If N is a name of an object class, we write the predicate $N(\_)$. We use an underscore to mark each place. Objects in S are treated as singleton-object classes and thus as object classes.
    Examples from the ORM in Fig. A.15 include $Person(\_)$, $SSN(\_)$, $Employee(\_)$, $Worker(\_)$, $Couple(\_)$, and $The\ Green\text{-}Grow\ Seed\ Company(\_)$. The ORM in Fig. A.16 has three object classes; each with one name. Thus, there are three object-class predicates: $A(\_)$, $B(\_)$, and $C(\_)$.

2. There are no predicates for dominant high-level object classes.

3. Except for aggregations and generalization-specializations, we map each n-ary ($n \geq 2$) relationship set of S to one n-place predicate for each of its names. In our examples here, we use the infix notation discussed in Chap. 2 for representing these predicates. We use the default *is member of* for an association without a name. Relationships in S are treated as singleton relationship sets having one and

only one relationship and having a 1:1 participation constraint for each of its connections.

In the ORM in Fig. A.15, for example, *Person (_) has Name (_)* is a two-place predicate derived from "*Person has Name*". Both *Employee(_) works for The Green-Grow Seed Company(_)* and *The Green-Grow Seed Company(_) employs Employee(_)* are two-place predicates derived from the binary relationship set between *Employee* and *The Green-Grow Seed Company*. Fig. A.16 has three relationship sets, each with one name, and thus we have three predicates: $A(\_) r_1 B(\_), B(\_) r_2 C(\_)$, and $B(\_) C(\_) A(\_)$.

4. The derivation of predicates for aggregation relationship sets depends on the aggregation hierarchy (or hierarchies) in which the aggregation relationship set is found. For each aggregation hierarchy in *S*, we derive a set of ordered pairs of object-class names as follows. Let *T* be a set of ordered pairs of object-class names, initially empty. Then, for each aggregation in the hierarchy with aggregate-object-class name(s) $C_1, \ldots, C_n$ and *m* subpart object classes whose names are $K_1, \ldots, K_p$ $(p \geq m)$, we add to *T* the pair (K_i, C_j) for $1 \leq i \leq n$ and $1 \leq j \leq p$. Let T^* be the transitive closure of *T*. We now derive a binary predicate $A(\_)$ *is subpart of* $B(\_)$ for each element *(A, B)* in T^*. Observe that we use only the default name *is subpart of*. Any user-provided name is ignored. Observe also that since we allow cycles in an aggregation hierarchy, *A* and *B* may be the same object-class name.

As an example, consider the *Order* aggregation hierarchy from the Green-Grow Seed Company in Fig. A.17. The transitive closure of object-class names (T^*) for this hierarchy is {*(Order, Order), (Item Description, Order), (Quantity, Item Description), (Seed Description, Item Description), (Quantity, Order), (Seed Description, Order)*}. Thus, the predicates we derive are as follows.

Order#(_) is subpart of Order(_)
Item Description(_) is subpart of Order(_)
Quantity(_) is subpart of Item Description(_)
Seed Description(_) is subpart of Item Description(_)
Quantity(_) is subpart of Order(_)
Seed Description(_) is subpart of Order(_)

5. There are no predicates for generalization-specialization relationship sets.

6. Since a relational object class is an object class, we map it to a one-place predicate for each of its names as explained in Case 1. In addition, each relational object class of *S* whose relationship set is *n*-ary maps to an $(n + 1)$-place predicate for each of its name combinations. We form a name combination for a relational object class by concatenating a name for the relational object class, a colon, a space, and a name of the relationship set in the relational object class. If there are *p* names for the relational object class and *q* names for the relationship set, there are *pq* name combinations.

In the ORM in Fig. A.15 we have one relational object class, *Couple* whose relationship set is the binary relationship set named both "*Employee has Spouse*" and "*Worker has Spouse*." We thus derive the two ternary predicates *Couple(_): Employee(_) has Spouse(_)* and *Couple(_): Worker(_) has Spouse(_)*.

A.1.2 Rules

Rules are derived from S as follows.

1. For each object class with multiple names $N_1, \ldots, N_n$, we write the rules:

$$\forall x\,[\,N_i\,(x) \Longleftrightarrow N_j\,(x)\,]$$

   for every i and j such that $1 \leq i \leq j \leq n$. Since these rules defines an equivalence class over the predicates for an object class, we are able to use any object-class name in place of another. Therefore, from here on we refer to *the* name of an object class to mean any one of its names.
   For example, we derive the rule $\forall x\,[Worker(x) \Longleftrightarrow Employee(x)]$ for the object class in the ORM in Fig. A.15 that has two names.

2. Except for aggregation and generalization-specialization relationship sets, we write equivalence-class rules for each n-ary relationship set with multiple names $N_1, \ldots, N_n$ as follows. Since the names do not necessarily reference the connections to the object classes in the same order, we are careful to associate the proper quantified variable with each connection in a relationship-set name. For an n-ary relationship-set name we will universally quantify the n variables $x_1, \ldots, x_n$ in the rules. To order these n variables in the rules properly, we associate each variable with a connection for the relationship set. We then let X_i represent these n variables written in their proper connection order for the relationship set named N_i, and write the rules

$$\forall x_1 \ldots \forall x_n\,[\,N_i(X_i) \Longleftrightarrow N_j(X_j)\,]$$

   for every i and j such that $1 \leq i < j \leq n$. Since these rules define an equivalence class over the predicates for a relationship set, we are able to use any relationship-set name in place of another. Therefore, from here on we refer to *the* name of a relationship set to mean any one of its names.
   For example, we derive the rule $\forall x \forall y\,[Employee(x)$ *works for The Green-Grow Seed Company*$(y) \Longleftrightarrow$ *The Green-Grow Seed Company*(y) *employs Employee*$(x)]$ for the relationship set in the ORM in Fig. A.15 that has two names. Here, we associate the universally quantified variable x with the connection to *Employee* and associate the universally quantified variable y with the connection to *The Green-Grow Seed Company*.

3. For each n-ary ($n \geq 2$) relationship set r we write a rule to ensure referential integrity. Let N_r be the name of r, and let C_i be the name of the object class connected to r in the ith position. We then write the rule

$$\forall x_1 \ldots \forall x_n\,\{N_r(x_1, \ldots x_n)\} \Rightarrow [C_1\,(x_1) \wedge \ldots \wedge C_n\,(x_n)]\}$$

   In the ORM in Fig. A.15, for example, we have $\forall x \forall y\,\{Person(x)$ *has Name*$(y) \Rightarrow [Person(x) \wedge Name(y)]\}$ for the "*Person has Name*" relationship set. Since we decompose aggregations into binary relationship sets and form their transitive closure, there are six referential-integrity rules for the aggregation hierarchy in Fig. A.17. For the *Quantity* object class in the transitively generated binary relationship set "*Quantity is subpart of Order*," for example, we have the referential-integrity rule: $\forall x \forall y \; \{Quantity(x) \; is \; subpart \; of \; Order(y) \Rightarrow [Quantity(x) \wedge Order(y)]\}$. For the ORM in Fig. A.16, we derive the following three rules.

$$\forall x \forall y \{A(x) \; r_1 \; B(y) \Rightarrow [A(x) \wedge B(y)]\}$$
$$\forall x \forall y \{B(x) \; r_1 \; C(y) \Rightarrow [B(x) \wedge C(y)]\}$$
$$\forall x \forall y \forall z \{B(x) \; C(y) \; A(z) \Rightarrow [B(x) \wedge C(y) \wedge A(z)]\}$$

4. For each relational object class we also write rules to ensure referential integrity. Let C be the name of the relational object class whose n-ary ($n \geq 2$) relationship set is r. Let N_r be the name of r. Let p be the $(n + 1)$-ary predicate generated for the relational object class as explained in Case 3 for predicates. We now write the rules

$$\forall x \{C(x) \Longleftrightarrow \exists x_1 \ldots \exists x_n \; [p(x, x_1, \ldots, x_n)]\}$$

$$\forall x_1 \ldots \forall x_n \{N_r(x_1, \ldots, x_n) \Longleftrightarrow \exists x \; [p(x, x_1, \ldots, x_n)]\}$$

For the relational object class in Fig. A.15, for example, we write the rules

$$\forall x \{Couple(x) \Longleftrightarrow \exists y \exists z \; [Couple(x): Employee(y) \; has \; Spouse(z)]\}$$

and

$$\forall x \forall y \; \{Employee(x) \; has \; Spouse(y)$$
$$\Longleftrightarrow \exists z \; [Couple(z): Employee(x) \; has \; Spouse(y)]\}$$

5. For each generalization-specialization, we write several rules. To give the rules, we first transform each generalization-specialization into a standard form by replacing any given participation constraint by a default participation constraint (1:1 for specialization object classes and 0:1 for a generalization object class). If in the replacement, we changed the participation constraint of a generalization object class from 1:1 to 0:1, we also impose a union specialization constraint. That is, we make the specialization constraint a union constraint if no specialization constraint was previously specified. We make the specialization constraint a partition constraint if the specialization constraint was previously a mutual-exclusion constraint, and we leave the specialization constraint unchanged otherwise. If in the replacement of participation constraints by their defaults, we made any other change, we write FALSE as a rule, which will cause any mathematical model instance to be invalid with respect to the ORM instance. Each generalization-specialization may have multiple generalizations and multiple specializations. Therefore, for each generalization-specialization, we let $G_1, \ldots, G_m$ be the names of the generalization object classes and let $S_1, \ldots, S_n$ be the names of the specialization object classes, and we write the following rule

$$\forall x \{[S_1(x) \vee \ldots \vee S_n(x)] \Rightarrow [G_1(x) \wedge \ldots \wedge G_m(x)]\}$$

If the generalization-specialization has a mutual-exclusion or partition constraint, we also include the rules

$$\forall x \; [S_i(x) \Rightarrow \neg S_j(x)]$$

for every i and j, $1 \leq i, j \leq n$ and $i \neq j$.

If the generalization-specialization has a union constraint or a partition constraint, it should have only one generalization. If not, we write FALSE as a rule. If so, we assume that the one generalization is named G, and write the rule

$$\forall x\{G(x) \Rightarrow [S_1(x) \vee \ldots \vee S_n(x)]\}$$

If the generalization-specialization has an intersection constraint, it should have only one specialization. If not, we write FALSE as a rule. If so, we assume that the one specialization is named S, and we write the rule

$$\forall x\{[G_1(x) \wedge \ldots \wedge G_m(x)] \Rightarrow S(x)\}$$

For the ORM in Fig. A.15 we have one generalization-specialization with a union constraint. We, therefore, write the following rules

$$\forall x\,[Employee(x) \Rightarrow Person(x)]$$
$$\forall x\,[Customer(x) \Rightarrow Person(x)]$$
$$\forall x\,\{Person(x) \Rightarrow [Employee(x) \vee Customer(x)]\}$$

We also have one generalization-specialization with no constraint. We, therefore, also write the rule
$$\forall x\,[Spouse(x) \Rightarrow Person(x)]$$

The astute reader will have noticed that these rules imply that every spouse is either an employee or a customer. To illustrate the union and keep the ORM small, we have done this intentionally.

6. For each aggregation hierarchy we wish to ensure that the relationship among the objects is asymmetric and transitive. Using the transitive closure T^* over the names of the object classes in the hierarchy developed in Case 4 of Sec. A.1.1, we write the rules as follows. To guarantee asymmetry, we write the rule

$$\forall x \forall y[A(x) \text{ is subpart of } B(y) \Rightarrow \neg\, A(y) \text{ is subpart of } B(x)]$$

for each element (A, B) in T^*. To guarantee transitivity, we write the rule

$$\forall x \forall y \forall z\{[A(x) \text{ is subpart of } B(y) \wedge B(y) \text{ is subpart of } C(z)]$$
$$\Rightarrow A(x) \text{ is subpart of } C(z)\}$$

for each triple of elements $(A, B)\ (B, C)\ (A, C)$ in T^*.

As an example, consider the *Order* hierarchy from the Green-Grow Seed Company in Fig. A.17. The following rules guarantee asymmetry.

$$\forall x \forall y[Order\#(x) \text{ is subpart of } Order(y)$$
$$\Rightarrow \neg\, Order\#(y) \text{ is subpart of } Order(x)]$$
$$\forall x \forall y[Item\ Description(x) \text{ is subpart of } Order(y)$$
$$\Rightarrow \neg\, Item\ Description(y) \text{ is subpart of } Order(x)]$$
$$\forall x \forall y[Quantity(x) \text{ is subpart of } Item\ Description(y)$$
$$\Rightarrow \neg\, Quantity(y) \text{ is subpart of } Item\ Description(x)]$$
$$\forall x \forall y[Seed\ Description(x) \text{ is subpart of } Item\ Description(y)$$
$$\Rightarrow \neg\, Seed\ Description(y) \text{ is subpart of } Item\ Description(x)]$$
$$\forall x \forall y[Quantity(x) \text{ is subpart of } Order(y)$$
$$\Rightarrow \neg\, Quantity(y) \text{ is subpart of } Order(x)]$$
$$\forall x \forall y[Seed\ Description(x) \text{ is subpart of } Order(y)$$
$$\Rightarrow \neg\, Seed\ Description(y) \text{ is subpart of } Order(x)]$$

The following rules guarantee transitivity.

$$\forall x \forall y \forall z \{[Quantity(x) \ is \ subpart \ of \ Item \ Description(y)$$
$$\wedge \ Item \ Description(y) \ is \ subpart \ of \ Order(z)]$$
$$\Rightarrow Quantity(x) \ is \ subpart \ of \ Order(z)\}$$
$$\forall x \forall y \forall z \{(Seed \ Description(x) \ is \ subpart \ of \ Item \ Description(y)$$
$$\wedge \ Item \ Description(y) \ is \ subpart \ of \ Order(z)]$$
$$\Rightarrow Seed \ Description(x) \ is \ subpart \ of \ Order(z)\}$$

7. For each participation constraint except those in generalization-specialization relationship sets, we write a rule to reflect the constraint. For aggregations, we consider each aggregation as if it were several binary relationship sets, one between each of its subpart object classes and its aggregate object class. To form participation-constraint rules, we let p be a participation constraint for a connection of an n-ary ($n \geq 2$) relationship set named N_r to an object class named C. Further, we let the connection be in the ith position for the n-place predicate. We assume that p has no shorthand notation and thus has the form $min_1 : max_1, \ldots,$ $min_m : max_m$. The rule we write has the form

$$t_1 \ \vee \ \ldots \ \vee \ t_m$$

where each t_i, $1 \leq i \leq m$, is a formula formed for $min_i : max_i$ as follows.

7a. If $min_i = 0$ and $max_i = \ ^*$, the term t_i is TRUE and we thus discard the rule. In the ORM in Figure A.15, for example, there is no rule for the $0 : ^*$ in the *"The Green-Grow Seed Company has Customer"* relationship set.

7b. If min_i and max_i are both positive integers, the term is

$$\forall x \{C(x) \Rightarrow min_i \leq \textit{Il}\!< x_1, \ldots, x_{i-1}, x_{i+1}, \ldots, x_m >$$
$$[N_r(x_1, \ldots, x_{i-1}, x, x_{i+1}, \ldots, x_m)]\}$$
$$\wedge$$
$$\forall x \{C(x) \Rightarrow \textit{Il}\!< x_1, \ldots, x_{i-1}, x_{i+1}, \ldots, x_m >$$
$$[N_r(x_1, \ldots, x_{i-1}, x, x_{i+1}, \ldots, x_m)] \leq max_i\}$$

Here, $< \ldots >$ is a tuple constructor, and $\textit{Il}x[P(x)]$ is the counting quantifier, which counts the number of different x-values for which the predicate P is true [Gries 1985]. The counting quantifier is first order if the universe is finite. Since our universes will be finite, we may use it.

An example from the ORM in Fig. A.15 is the participation constraint 1 ($= 1{:}1$) in the *"Person has Name"* relationship set. We write this constraint as

$$\forall x \{Person(x) \Rightarrow 1 \leq \textit{Il}\!<y>[Person(x) \ has \ Name(y)]\}$$
$$\wedge$$
$$\forall x \{Person(x) \Rightarrow \textit{Il}\!<y>(Person(x) \ has \ Name(y)] \leq 1\}$$

7c. If min_i is a positive integer and max_i is $\ ^*$, the term is

$$\forall x \{C(x) \Rightarrow min_i \leq \textit{Il}\!<x_1, \ldots, x_{i-1}, x_{i+1}, \ldots, x_m >$$
$$[N_r(x_1, \ldots, x_{i-1}, x, x_{i+1}, \ldots, x_m)]\}$$

which is the first conjunct of the term for Case 7b.

In the ORM in Fig. A.15, for example, we write the $1:*$ constraint in the "*Person has Name*" relationship set as

$$\forall x\{Name(x) \Longrightarrow 1 \le \text{\textnormal{\Mobj}}< y>[Person(y) \text{ has } Name(x)]\}$$

7d. If $min_i = 0$ and max_i is a positive integer, we write

$$\forall x\{C(x) \Longrightarrow \text{\Mobj}<x_1, \ldots, x_{i-1}, x_{i+1}, \ldots, x_m>$$
$$[N_r(x_1, \ldots, x_{i-1}, x, x_{i+1}, \ldots, x_m)] \le max_i)$$

which is the last conjunct of the term for Case 7b.
In the ORM in Fig. A.15, for example, we write the $0:1$ constraint in the "*Employee has Spouse*" relationship set as

$$\forall x\{Employee(x) \Longrightarrow \text{\Mobj}<y>[Employee(x) \text{ has } Spouse(y)] \le 1\}$$

7e. If either min_i or max_i includes a variable or a function, we first decide for each variable whether it is Type 1 or Type 2. A variable is *Type 1* if it is included in the participation constraints for one and only one object class, and is *Type 2* otherwise. In Fig. A.16, for example, a is a Type 1 variable because it is included in participation constraints only for object class B. Although a is included in other constraints, it is not included in a participation constraint for any other object class. On the other hand, b, which is mentioned in the participation constraints for both object classes A and B, is Type 2. We assign each Type 1 variable v the Skolem function f_v. (If there are non-Skolem function names of this form, we have to change their names. In Fig. A.16 neither of the two non-Skolem functions, addition in $a + b$ and multiplication in $2a$, has this form, so no function-name changes are necessary.)
With this preparation, we now write the term for $min_i : max_i$ as

$$\forall x\{C(x) \Longrightarrow min_i / f \le \text{\Mobj}<x_1, \ldots, x_{i-1}, x_{i+1}, \ldots, x_m>$$
$$[N_r(x_1, \ldots, x_{i-1}, x, x_{i+1}, \ldots, x_m)]\}$$
$$\wedge$$
$$\forall x\{C(x) \Longrightarrow \text{\Mobj}<x_1, \ldots, x_{i-1}, x_{i+1}, \ldots, x_m>$$
$$[N_r(x_1, \ldots, x_{i-1}, x, x_{i+1}, \ldots, x_m)] \le max_i / f\}$$

where min_i / f and max_i / f are respectively min_i and max_i with each Type 1 variable replaced by its assigned Skolem function using the universally quantified variable x as its argument. As before, we may omit the first conjunct of the term if min_i is zero, and we may omit the last conjunct of the term if max_i is $*$. The Type 2 participation-constraint variables remain as they are and become constants in the first-order rules we are deriving. The functions also remain as they are.
For the ORM in Fig. A.16, for example, the derived participation-constraint rules are

$$\forall x\{A(x) \Longrightarrow \text{\Mobj}<y>[r_1(x, y)] \le b\}$$
$$\forall x\{B(x) \Longrightarrow f_a(x) \le \text{\Mobj}<y>[r_1(y, x)]\}$$
$$\forall x\{B(x) \Longrightarrow \text{\Mobj}<y>[r_1(y, x)] \le f_a(x)+b\}$$
$$\forall x\{B(x) \Longrightarrow 2f_a(x) \le \text{\Mobj}<y>[r_2(x, y)]\}$$

where f_a is the Skolem function assigned to a.

8. For each co-occurrence constraint we write a rule as follows. Let c be a co-occurrence constraint for relationship set r. Let N_r be the name of r. Let c have n object classes of r listed for its left-hand side and m object classes of r listed for its right-hand side, and let p be the number of object classes of r not listed in either the left-hand side or the right-hand side. We assume that the constraint for c has no shorthand notation and thus is of the form $min_1 : max_1, \ldots min_q : max_q$. We also assume, without loss of generality, that the order of the arguments for r are the n left-hand-side object classes, followed by the m right-hand-side object classes, followed by the p unlisted object classes.

The rule we write for a co-occurrence constraint depends on whether the co-occurrence constraint has any Type 1 variables. If there are no Type 1 variables, the rule has the form

$$t_1 \vee \ldots \vee t_q$$

where each t_i, $1 \le i \le q$, is a formula formed for $min_i : max_i$ as follows:

$$\forall x_1 \ldots \forall x_n$$
$$[\exists v_1 \ldots \exists v_m \exists w_1 \ldots \exists w_p (N_r(x_1, \ldots, x_n, v_1, \ldots, v_m, w_1, \ldots, w_p)$$
$$\Rightarrow$$
$$[\![min_i \le |\!|\!\!<y_1, \ldots, y_m>$$
$$\{\exists z_1 \ldots \exists z_p [N_r(x_1, \ldots, x_n, y_1, \ldots, y_m, z_1, \ldots, z_p)]\}]\!|]\!]$$
$$\wedge$$
$$\forall x_1 \ldots \forall x_n$$
$$[\exists v_1 \ldots \exists v_m \exists w_1 \ldots \exists w_p [(N_r(x_1, \ldots, x_n, v_1, \ldots, v_m, w_1, \ldots, w_p)$$
$$\Rightarrow$$
$$[\![|\!|\!\!<y_1, \ldots, y_m>$$
$$\{\exists z_1 \ldots \exists z_p [N_r(x_1, \ldots, x_n, y_1, \ldots, y_m, z_1, \ldots, z_p)]\} \le max_i]\!|]\!]$$

If there are s Type 1 variables, $s \ge 1$, associated with object classes named $C_1, \ldots, C_s$, the rule has the form

$$\forall z_1 \ldots \forall z_s \{ [C_1(z_1) \wedge \ldots \wedge C_s(z_s)] \Rightarrow (t_1 \vee \ldots \vee t_q) \}$$

where each t_i, $1 \le i \le q$, is a formula formed for $min_i : max_i$ as before except that min_i is replaced by min_i / f and max_i is replaced by max_i / f. Here, as in Case 7, $/f$ means that we substitute Skolem functions for Type 1 variables. We must also be sure to properly match the universally-quantified variables for the Skolem functions and the object-class names. In both cases (with or without Skolem functions), we may omit the last conjunct if max_i is $\star$. We do not omit the first conjunct, however, because min_i cannot be zero.

For the co-occurrence constraint in Fig. A.16, we write the rule

$$\{ \forall x [\exists v \exists w (B(x)C(v)A(w)$$
$$\Rightarrow [\![1 \le |\!|\!\!<y>\{\exists z[B(x)C(y)A(z)]\}]\!|]\!)]$$
$$\wedge$$
$$\forall x [\exists v \exists w (B(x)C(v)A(w)$$
$$\Rightarrow [\![|\!|\!\!<y>\{\exists z[B(x)C(y)A(z)]\} \le 1]\!|]\!)]\}$$
$$\vee$$
$$\forall x [\exists v \exists w (B(x)C(v)A(w)$$
$$\Rightarrow [\![b \le |\!|\!\!<y>\{\exists z[B(x)C(y)A(z)]\}]\!|]\!)]$$

If in the co-occurrence constraint in Fig. A.16 we replace *1:1, b:* * by *a:b*, we would instead write the rule

$$\forall z[\![B(z) \Rightarrow \{\forall x[\exists v \exists w(B(x)C(v)A(w)$$
$$\Rightarrow [\![f_a(z) \le \mathsf{N}{<}y{>}\{\exists z[B(x)C(y)A(z)]\}]\!])]$$
$$\wedge$$
$$\forall x[\exists v \exists w(B(x)C(v)A(w)$$
$$\Rightarrow [\![\mathsf{N}{<}y{>}\{\exists z[B(x)C(y)A(z)]\} \le b]\!])]\}]\!]$$

9. For each object-class cardinality constraint we write a rule as follows. Let C be the name of an object class with a cardinality constraint. We assume that there is no shorthand notation and thus that the constraint for C has the form $min_1 : max_1$, $\ldots min_n : max_n$. As for co-occurrence constraints, the rule we write depends on whether the object-class cardinality constraint has any Type 1 variables. If there are no Type 1 variables, the rule has the form

$$t_1 \vee \ldots \vee t_n$$

where each t_i, $1 \le i \le n$, is a formula formed for $min_i : max_i$ as follows:

$$min_i \le \mathsf{N}x[C(x)] \wedge \mathsf{N}x[C(x)] \le max_i$$

If there are m Type 1 variables, $m \ge 1$, associated with object classes named C_1, $\ldots, C_m$, the rule has the form

$$\forall y_1 \ldots \forall y_m\{[C_1(y_1) \wedge \ldots \wedge C_m(y_m)] \Rightarrow (t_1 \vee \ldots \vee t_n)\}$$

where each t_i, $1 \le i \le n$, is a formula formed for $min_i : max_i$ as before except that min_i is replaced by min_i / f and max_i is replaced by max_i / f. Here, as in Case 7, $/f$ means that we substitute Skolem functions for Type 1 variables. We must again be sure to properly match the universally quantified variables for the Skolem functions and the object-class names. In both cases (with or without Skolem functions), we may omit the first conjunct if min_i is 0, and we may omit the last conjunct if max_i is *.

In Fig. A.16, object class C has the object-class cardinality constraint *4:8*. We, therefore, write the rule

$$4 \le \mathsf{N}x[C(x)] \wedge \mathsf{N}x[C(x)] \le 8$$

10. For each high-level relationship set r whose construction is not explicitly given, and thus by default is the join, we write a rule as follows. Let N_r be the name of an n-ary high-level relationship set connected to object classes named $C_1, \ldots, C_n$. Let $K_1, \ldots, K_m$ be the names of the object classes included in r, and let $N_{r_1}, \ldots, N_{r_p}$ be the names of the relationship sets included in r. If any of these relationship sets that are aggregations, we use their binary decompositions instead. We then write the rule

$$\forall x_1 \ldots \forall x_n \forall y_1 \ldots \forall y_m\{[C_1(x_1) \wedge \ldots \wedge C_n(x_n)$$
$$\wedge K_1(y_1) \wedge \ldots \wedge K_m(y_m) \wedge N_{r_1}(Z_1) \wedge \ldots N_{r_p}(Z_p)]$$
$$\Rightarrow N_r(x_1, \ldots, x_n)\}$$

where, without loss of generality, we may assume that the order of the arguments in r is consistent with the placement of the associated object classes, and where Z_i, $1 \leq i \leq p$, represents the proper selection of the appropriate number of bound variables from among $x_1, \ldots, x_n, y_1, \ldots, y_m$ for each relationship set.

Assume, for example, that in the ORM in Fig. A.15 we create a high-level relationship set "*Employee serves Customer*" that includes the relationship sets "*Employee works for The Green-Grow Seed Company*," "*The Green-Grow Seed Company has Customer*," and "*The Green-Grow Seed Company is located on Rural Route #1*," and that includes the object classes *The Green-Grow Seed Company* and *Rural Route #1*. We would then include the rule

$$\forall x_1 \forall x_2 \forall y_1 \forall y_2$$
$$\{[Employee(x_1) \wedge Customer(x_2)$$
$$\wedge \text{ } The \text{ } Green\text{-}Grow \text{ } Seed \text{ } Company(y_1) \wedge Rural \text{ } Route \text{ } \#1 \text{ } (y_2)$$
$$\wedge \text{ } Employee(x_1) \text{ } works \text{ } for \text{ } The \text{ } Green\text{-}Grow \text{ } Seed \text{ } Company(y_1)$$
$$\wedge \text{ } The \text{ } Green\text{-}Grow \text{ } Seed \text{ } Company(y_1) \text{ } has \text{ } Customer(x_2)$$
$$\wedge \text{ } The \text{ } Green\text{-}Grow \text{ } Seed \text{ } Company(y_1) \text{ } is \text{ } located \text{ } on$$
$$\qquad Rural \text{ } Route \text{ } \#1 \text{ } (y_2)]$$
$$\Rightarrow Employee(x_1) \text{ } serves \text{ } Customer(x_2)\}$$

11. Any general constraint recognized as a first-order statement is included as a rule. If there are no Type 1 variables, we include it directly. If there are n Type 1 variables, $n \geq 1$, associated with object classes named $C_1, \ldots, C_n$, and if g is the general constraint, we write the rule

$$\forall x_1 \ldots \forall x_n \{[C_1(x_1) \wedge \ldots \wedge C_n(x_n)] \Rightarrow g/f\}$$

where g/f means that we substitute Skolem functions for Type 1 variables. We must again be sure to properly match the universally quantified variables for the Skolem functions and the object-class names.

As examples, in Fig. A.16 we have three general constraints, $2 \leq a$, $a \leq 3$, and $b = |B|$. Since a is a Type 1 variable that is associated with the Skolem function f_a and the object class B, we include the following rules:

$$\forall x \{B(x) \Rightarrow [2 \leq f_a(x)]\}$$
$$\forall x \{B(x) \Rightarrow [f_a(x) \leq 3]\}$$

The constraint $b = |B|$ is an alternate syntax for

$$b = \mathbb{N}x[B(x)]$$

We, therefore, include it as a rule in this form.

12. As we conclude our section on rules, we remark that, although association is a special type of relationship in OSA, it does not impose any special constraints. An association construct provides us with a default name *is member of* for the relationship set and draws our attention to the set of elements in the member class that are connected to an element in the set class. Since this set is immediately derivable for every element in the set class, we need no constraints and thus no rules.

A.2 Mapping from the First-Order Language to a Mathematical Model

When we complete a translation for an OSA model instance, the result is a first-order language L and a set of rules Z. We give an interpretation for L by mapping it to a mathematical model instance M.

A mathematical model M is a quadruple (U, R, C, F) where U is a finite universe of objects, R is a finite set of unary, binary, . . . , n-ary relations over U, C is a set of constants chosen from U, and F is a finite set of one-place functions from U to U, two-place functions from U^2 to U, . . . , m-place functions from U^n to U. Here, m and n are finite, but arbitrary.

We interpret L by establishing a mapping from L to M as follows. We map the quantifiers ($\forall$, $\exists$, $\mathbb{N}$) to U, which means that when we evaluate them, the set of objects we consider is the set of objects in U. We map each n-place predicate p in L to the n-ary relation p' in R. We map each constant in L to an element of C. We map each n-place function f in L to the n-place function g in F.

Once we establish the mapping, we can check the truth of Z with respect to M by evaluating the rules in Z in terms of the mapping. If all the rules hold, M is valid. Figure A.18 shows a valid model instance for our ORM in Fig. A.15, and Fig. A.19 shows a valid model instance for our ORM in Fig. A.16.

A.3 The OSA Meta-Model

To complete our formal definition for valid OSA model instances, we now need the OSA meta-model in Figs. A.1 through A.14. From the meta-model, we can derive predicates and rules as we have described. We can then provide a mathematical model instance and establish an interpretation. If all the rules derived from our meta-model hold for the mathematical model instance, we say that the mathematical model instance is a *valid OSA model instance*.

As an example, we give in Fig. A.20, a mathematical model instance for the ORM in Fig. A.16. Each object class, relationship set, and constraint is captured in an appropriate unary relation, and the associations among the object classes, relationship sets, and constraints are captured in appropriate n-ary relations. (In Fig. A.20 we have omitted relations of the mathematical model instance that are empty.) The derived predicates and rules for the ORM in Fig. A.16 have all been given as examples in Sec. A.1 of this appendix. They all hold in the mathematical model instance in Fig. A.20. Since all the rules hold, the OSA model instance in Fig. A.16 is a valid OSA model instance.

The general constraints in our meta-model are written in natural language so that the meta-model diagrams would not be encumbered with predicate-calculus syntax. To fully translate the meta-model into predicates and rules as we have described, we must provide first-order formulas for each of the general constraints in our meta-model. As examples, we give translations below for the first four general constraints in Fig. A.4. We leave the translation of the remaining general constraints as exercises. So that our formal definition is complete, however, we provide these definitions in the answers to exercises [Embley 1991a], and thus provide a complete, formal definition for valid OSA model instances.

A.3.1 Translation of the First Constraint in Fig. A.4

Constraint

> *A Dominant High-Level Object Class must have the same Object Class Name(s) as the Object Class it dominates.*

Translation

$$\forall x_1 \forall x_2 [\![Dominant\ High\text{-}Level\ Object\ Class(x_1)\ dominates\ Object\ Class(x_2)$$
$$\Rightarrow \{\forall x_3 [Object\ Class\ Name(x_3)\ names\ Object\ Class(x_1)$$
$$\Longleftrightarrow Object\ Class\ Name(x_3)\ names\ Object\ Class(x_2)]\}]\!]$$

A.3.2 Translation of the Second Constraint in Fig. A.4

Constraint

> *The components of any High-Level Object Class, High-Level Relationship Set, or Relational Object Class included in High-Level Object Class C must also be included in C.*

Translation

$\forall x_1 (High\text{-}Level\ Object\ Class(x_1) \Rightarrow$
$\quad ([\![\forall x_2 \forall x_3 \{[High\text{-}Level\ Object\ Class(x_1)\ includes\ Object\ Class(x_2)$
$\quad\quad \wedge\ High\text{-}Level\ Object\ Class(x_2)\ includes\ Object\ Class(x_3)]$
$\quad\quad \Rightarrow High\text{-}Level\ Object\ Class(x_1)\ includes\ Object\ Class(x_3)\}]\!]$

$\wedge$
$[\![\forall x_2 \forall x_3 \{[High\text{-}Level\ Object\ Class(x_1)\ includes\ Object\ Class(x_2)$
$\quad \wedge\ High\text{-}Level\ Object\ Class(x_2)\ includes\ Relationship\ Set(x_3)]$
$\quad \Rightarrow High\text{-}Level\ Object\ Class(x_1)\ includes\ Relationship\ Set(x_3)\}]\!]$

$\wedge$
$[\![\forall x_2 \forall x_3 \{[High\text{-}Level\ Object\ Class(x_1)\ includes\ Object\ Class(x_2)$
$\quad \wedge\ High\text{-}Level\ Object\ Class(x_2)\ includes\ Constraint(x_3)]$
$\quad \Rightarrow High\text{-}Level\ Object\ Class(x_1)\ includes\ Constraint(x_3)\}]\!]$

$\wedge$
$[\![\forall x_2 \forall x_3 \{[High\text{-}Level\ Object\ Class(x_1)\ includes\ Object\ Class(x_2)$
$\quad \wedge\ High\text{-}Level\ Object\ Class(x_2)\ includes\ Note(x_3)]$
$\quad \Rightarrow High\text{-}Level\ Object\ Class(x_1)\ includes\ Note(x_3)\}]\!]$

$\wedge$
$[\![\forall x_2 \forall x_3 \{[High\text{-}Level\ Object\ Class(x_1)\ includes\ Relationship\ Set(x_2)$
$\quad \wedge\ High\text{-}Level\ Relationship\ Set(x_2)\ includes\ Object\ Class(x_3)]$
$\quad \Rightarrow High\text{-}Level\ Object\ Class(x_1)\ includes\ Object\ Class(x_3)\}]\!]$

$\wedge$
$[\![\forall x_2 \forall x_3 \{[High\text{-}Level\ Object\ Class(x_1)\ includes\ Relationship\ Set(x_2)$
$\quad \wedge\ High\text{-}Level\ Relationship\ Set(x_2)\ includes\ Relationship\ Set(x_3)]$
$\quad \Rightarrow High\text{-}Level\ Object\ Class(x_1)\ includes\ Relationship\ Set(x_3)\}]\!]$

$\wedge$
$[\![\forall x_2 \forall x_3 \{[High\text{-}Level\ Object\ Class(x_1)\ includes\ Relationship\ Set(x_2)$
$\quad \wedge\ High\text{-}Level\ Relationship\ Set(x_2)\ includes\ Constraint(x_3)]$
$\quad \Rightarrow High\text{-}Level\ Object\ Class(x_1)\ includes\ Constraint(x_3)\}]\!]$

$$\overset{\wedge}{[\![}\forall x_2 \forall x_3 \{[\textit{High-Level Object Class}(x_1) \textit{ includes Relationship Set}(x_2)$$
$$\wedge\ \textit{High-Level Relationship Set}(x_2) \textit{ includes Note}(x_3)]$$
$$\Rightarrow \textit{High-Level Object Class}(x_1) \textit{ includes Note}(x_3)\}]\!]$$

$$\overset{\wedge}{[\![}\forall x_2 \forall x_3 \forall x_4 \forall x_5 \{[\textit{High-Level Object Class}(x_1) \textit{ includes Object Class}(x_2)$$
$$\wedge\ \textit{Relational Object Class}(x_2) \textit{ consists of Relationship Set}(x_3)$$
$$\wedge\ \textit{Object Class}(x_4) \textit{ has connection to Relationship Set}(x_3)]$$
$$\Rightarrow \textit{High-Level Object Class}(x_1) \textit{ includes Object Class}(x_4)\}]\!]$$

$$\overset{\wedge}{[\![}\forall x_2 \forall x_3 \{[\textit{High-Level Object Class}(x_1) \textit{ includes Object Class}(x_2)$$
$$\wedge\ \textit{Relational Object Class}(x_2) \textit{ consists of Relationship Set}(x_3)]$$
$$\Rightarrow \textit{High-Level Object Class}(x_1) \textit{ includes Relationship Set}(x_3)]\}]\!])$$

A.3.3 Translation of the Third Constraint in Fig. A.4

Constraint

A High-Level Object Class must not include itself.

Translation

$$\neg\ \exists x_1 [\textit{High-Level Object Class}(x_1) \textit{ includes Object Class}(x_1)]$$

A.3.4 Translation of the Fourth Constraint in Fig. A.4

Constraint

If a High-Level Object Class includes Relationship Set R, it must also include all the Object Classes connected to R.

Translation

$$\forall x_1 \forall x_2 [\![\textit{High-Level Object Class}(x_1) \textit{ includes Relationship Set}(x_2)$$
$$\Rightarrow \{\forall x_3 \forall x_4 [\textit{Object Class}(x_3) \textit{ has Connection}(x_4) \textit{ to Relationship Set}(x_2)$$
$$\Rightarrow \textit{High-Level Object Class}(x_1) \textit{ includes Object Class}(x_3)]\}]\!]$$

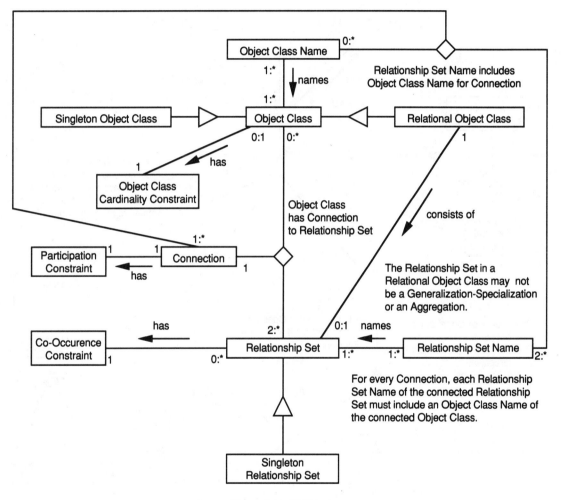

Figure A.1   ORM basics.

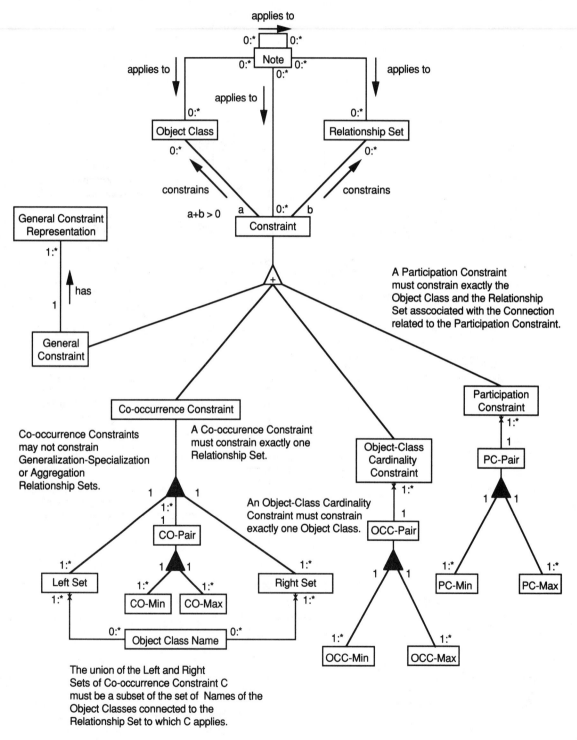

Figure A.2   ORM notes and cardinality constraints.

A Specialization Constraint must constrain exactly one Relationship Set.

If a Generalization-Specialization has a Partition or Union Constraint, the Generalization-Specialization has only one Generalization Class.

A Partition Constraint must constrain all its connected Object Classes and no others.

A Mutual Exclusion Constraint must constrain all its Specialization Classes and no others.

A Union Constraint must constrain its connected Generalization Class and no others.

If a Generalization-Specialization has an Intersection Constraint, the Generalization-Specialization has only one Specialization Class.

An Intersection Constraint must constrain exactly one Relationship Set and must constrain its Specialization Class and no others.

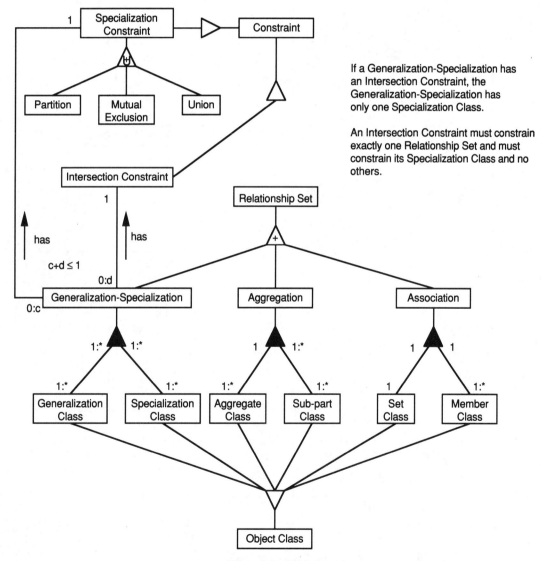

Figure A.3   ORM abstractions.

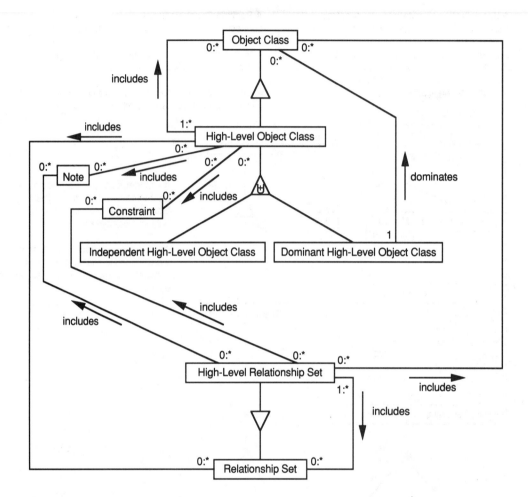

A Dominant High-Level Object Class must have the same Object Class Name(s) as the Object Class it dominates.

The components of any High-Level Object Class, High-Level Relationship Set, or Relational Object Class included in High-Level Object Class C must also be included in C.

A High-Level Object Class must not include itself.

If a High-Level Object Class includes Relationship Set R, it must also include all the Object Classes connected to R.

The components of any High-Level Object Class, High-Level Relationship Set, or Relational Object Class included in High-Level Relationship Set R must also be included in R.

A High-Level Relationship Set must not include itself.

If a High-Level Relationship Set includes Object Class C, it must also include all the Relationship Sets connected to C.

Figure A.4   ORM high-level object classes and relationship sets.

A State and Transition can be connected
only if the State and Transition are part
of the same State Net.

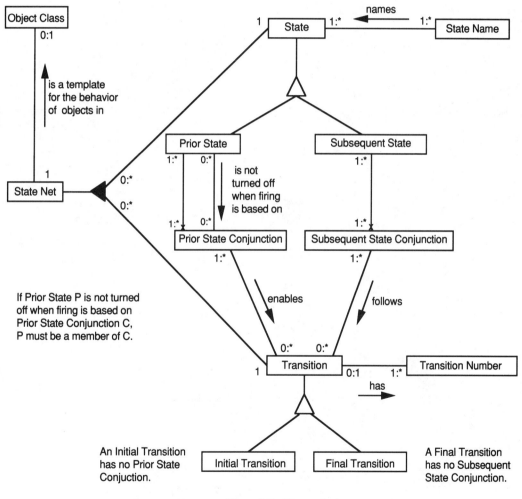

Figure A.5   State net basics.

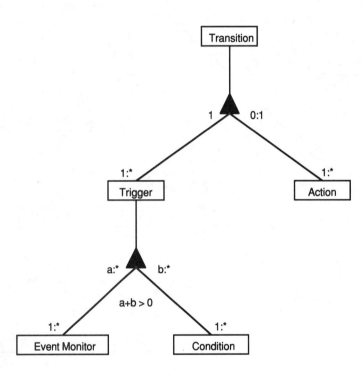

Figure A.6   State net transition details.

A Real-Time Duration Constraint can only constrain
Real-Time Duration Constrainable Components.

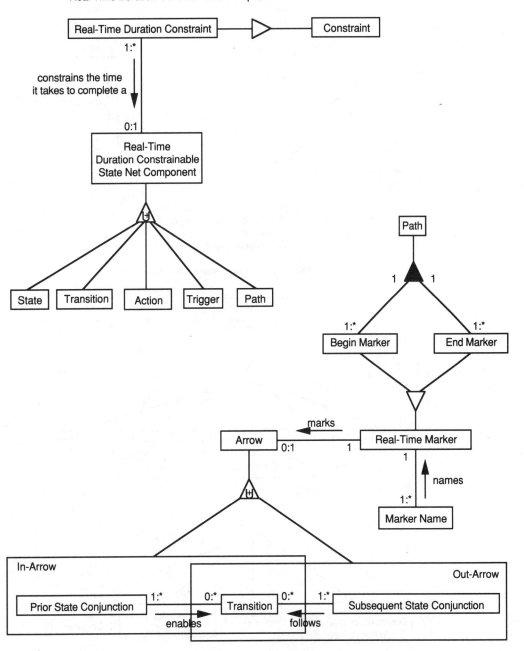

Figure A.7   State net real-time details.

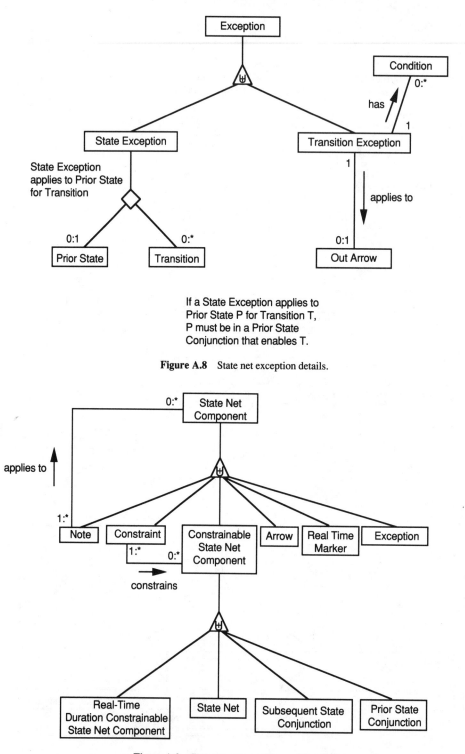

Figure A.8   State net exception details.

Figure A.9   State net notes and general constraints.

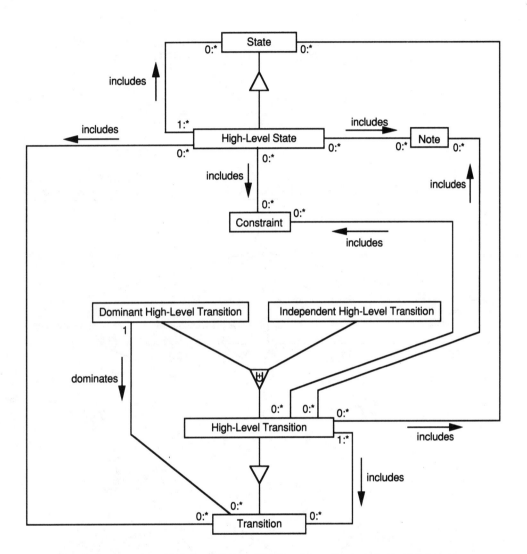

The components of any High-Level State or High-Level Transition included in High-Level State S must also be included in S.

A High-Level State must not include itself.

If a high-Level State includes Transition T, it must also include all the States connected to T.

The components of any High-Level State or High-Level Transition included in High-Level Transition T must also be included in T.

A High-Level Transition must not include itself.

If a High-Level Transition includes State S, it must also include all the Transitions connected to S.

A Dominant High-Level Transition must have the same Trigger and Action as the Transition it dominates.

Figure A.10   State net high-level state and transition details.

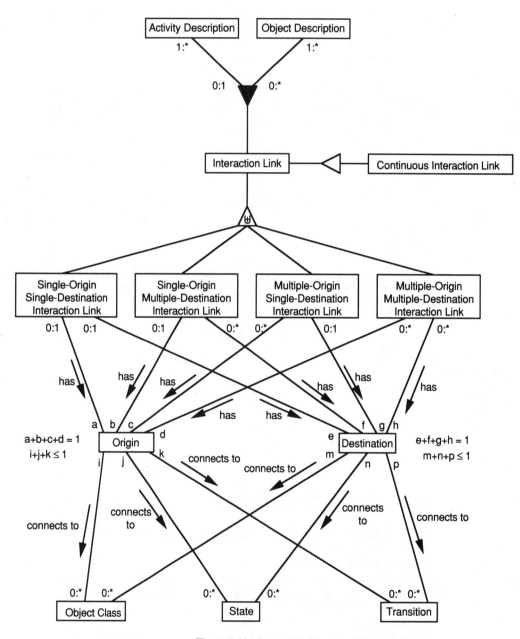

Figure A.11 Interaction diagram basics.

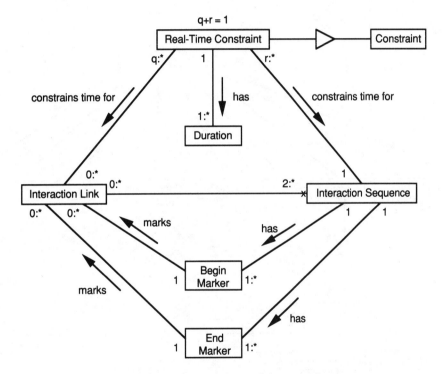

Figure A.12   Interaction diagram real-time constraints.

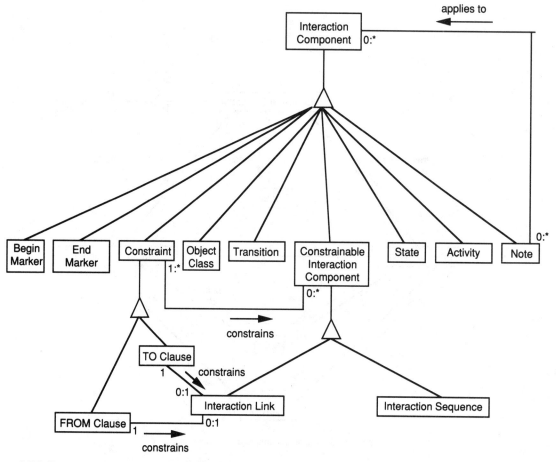

A TO Clause can only
constrain an Interaction Link.

A FROM Clause can only
constrain an Interaction Link.

Figure A.13   Interaction diagram notes and constraints.

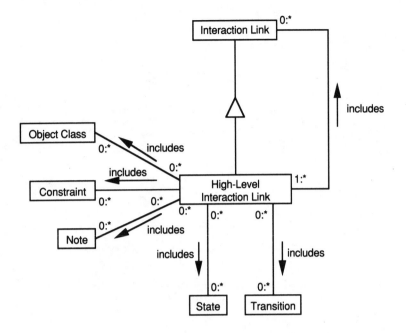

The components of any High-Level Interaction Link included in High-Level Interaction Link L must also be included in L.

A High-Level Interaction Link must not include itself.

Figure A.14   Interaction diagram high-level interaction links.

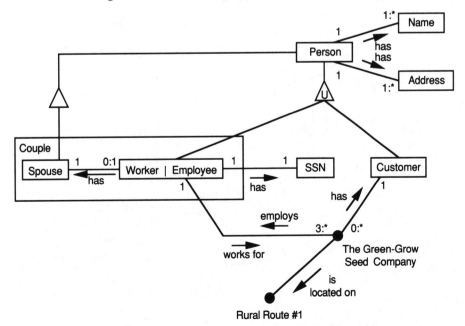

Figure A.15   ORM for a small part of the Green-Grow Seed Company.

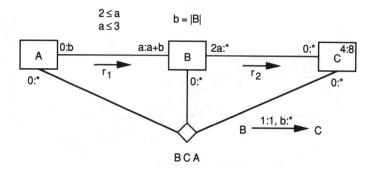

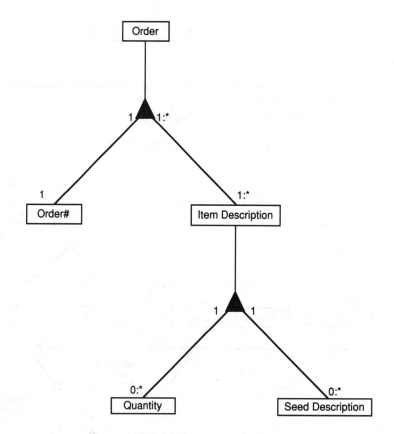

Figure A.16   An abstract ORM.

Figure A.17   Subparts of an order.

U = { Person1, Person2, Person3, Person4, Couple1, Mary, John, Zed, 10 State Street, 12 Maple, 17 Elm,
       Rural Route #1,111-11-1111, 222-22-2222, 333-33-3333, The Green-Grow Seed Company}

R = {

Person'	Name'	Address'	Customer'	SSN'	Person has Name'	
Person1	Mary	10 State Street	Person1	111-11-1111	Person1	John
Person2	John	12 Maple	Person2	222-22-2222	Person2	Zed
Person3	Zed	17 Elm	Person3	333-33-3333	Person3	Mary
Person4					Person4	John

The Green-Grow Seed Company'	The Green-Grow Seed Company is located on Rural Route #1'	
The Green-Grow Seed Company	The Green-Grow Seed Company	Rural Route #1

Rural Route #1'	Spouse'	Couple'	The Green-Grow Seed Company has Customer'	
Rural Route #1	Person1	Couple1	The Green-Grow Seed Company	Person1
			The Green-Grow Seed Company	Person2
			The Green-Grow Seed Company	Person3

Employee'	Worker'	Employee works for The Green-Grow Seed Company'	
Person2	Person2	Person2	The Green-Grow Seed Company
Person3	Person3	Person3	The Green-Grow Seed Company
Person4	Person4	Person4	The Green-Grow Seed Company

Person has Address'		Worker works for The Green-Grow Seed Company'	
Person1	10 State Street	Person2	The Green-Grow Seed Company
Person2	12 Maple	Person3	The Green-Grow Seed Company
Person3	10 State Street	Person4	The Green-Grow Seed Company
Person4	17 Elm		

Employee has SSN'		The Green-Grow Seed Company employs Employee'	
Person2	111-11-1111	The Green-Grow Seed Company	Person2
Person3	222-22-2222	The Green-Grow Seed Company	Person3
Person4	333-33-3333	The Green-Grow Seed Company	Person4

Worker has SSN'		The Green-Grow Seed Company employs Worker'	
Person2	111-11-1111	The Green-Grow Seed Company	Person2
Person3	222-22-2222	The Green-Grow Seed Company	Person3
Person4	333-33-3333	The Green-Grow Seed Company	Person4

Couple: Employee has Spouse'			Couple: Worker has Spouse'		
Couple1	Person3	Person1	Couple1	Person3	Person1

Worker has Spouse		Employee has Spouse	
Person3	Person1	Person3	Person1

}

C = { }

F = { }

Figure A.18 A mathematical model instance for the ORM in Fig. A.15.

U = {1, 2, 3, 4, 5, 6, ?}

R = {

A'	B'	C'	A r₁ B'	B r₂ C'	B C A'
1	1	1	1 1	1 1	1 1 1
2	2	2	2 1	1 2	2 1 2
3	3	3	1 2	1 3	2 2 3
4		4	2 2	1 4	2 3 4
5		5	3 2	2 1	3 1 5
6		6	1 3	2 2	3 2 6
			2 3	2 3	3 3 1
			3 3	2 4	3 4 2
			4 3	2 5	
			5 3	2 6	
				3 1	
				3 2	
				3 3	
				3 4	}

C = { b = 3 }

F = { $f_a^{'}$ = {(1, 2), (2, 3), (3, 2), (4, ?), (5, ?), (6, ?)},

   addition   = {((1, 1), 2), ((1, 2), 3), ((1, 3), 4), ((1, 4), 5), ((1, 5), 6), ((1, 6), ?), ((1, ?), ?),
                 ((2, 1), 3), ((2, 2), 4), ((2, 3), 5), ((2, 4), 6), ((2, 5), ?), ((2, 6), ?), ((2, ?), ?),
                 ((3, 1), 4), ((3, 2), 5), ((3, 3), 6), ((3, 4), ?), ((3, 5), ?), ((3, 6), ?), ((3, ?), ?),
                 ((4, 1), 5), ((4, 2), 6), ((4, 3), ?), ((4, 4), ?), ((4, 5), ?), ((4, 6), ?), ((4, ?), ?),
                 ((5, 1), 6), ((5, 2), ?), ((5, 3), ?), ((5, 4), ?), ((5, 5), ?), ((5, 6), ?), ((5, ?), ?),
                 ((6, 1), ?), ((6, 2), ?), ((6, 3), ?), ((6, 4), ?), ((6, 5), ?), ((6, 6), ?), ((6, ?), ?),
                 ((?, 1), ?), ((?, 2), ?), ((?, 3), ?), ((?, 4), ?), ((?, 5), ?), ((?, 6), ?), ((?, ?), ?)},

   multiplication = {((1, 1), 1), ((1, 2), 2), ((1, 3), 3), ((1, 4), 4), ((1, 5), 5), ((1, 6), 6), ((1, ?) ?),
                 ((2, 1), 2), ((2, 2), 4), ((2, 3), 6), ((2, 4), ?), ((2, 5), ?), ((2, 6), ?), ((2, ?), ?),
                 ((3, 1), 3), ((3, 2), 6), ((3, 3), ?), ((3, 4), ?), ((3, 5), ?), ((3, 6), ?), ((3, ?), ?),
                 ((4, 1), 4), ((4, 2), ?), ((4, 3), ?), ((4, 4), ?), ((4, 5), ?), ((4, 6), ?), ((4, ?), ?),
                 ((5, 1), 5), ((5, 2), ?), ((5, 3), ?), ((5, 4), ?), ((5, 5), ?), ((5, 6), ?), ((5, ?), ?),
                 ((6, 1), 6), ((6, 2), ?), ((6, 3), ?), ((6, 4), ?), ((6, 5), ?), ((6, 6), ?), ((6, ?), ?),
                 ((?, 1), ?), ((?, 2), ?), ((?, 3), ?), ((?, 4), ?), ((?, 5), ?), ((?, 6), ?), ((?, ?), ?)}

}

Figure A.19   A mathematical model instance for the ORM in Fig. A.16.

U = "all the symbols listed in the relations below"   C = { }   F = { }

R = {Object Class'

| A |
| B |
| C |

Object Class Name'

A
B
C

Relationship Set'

A	r_1	B
B	r_2	C
B	C	A

Relationship Set Name'

A r_1 B
B r_2 C
B C A

Connection'

Connection1
Connection2
Connection3
Connection4
Connection5
Connection6
Connection7

Constraint'

Constraint1
Constraint2
Constraint3
Constraint4
Constraint5
Constraint6
Constraint7
Constraint8
Constraint9
Constraint10
Constraint11
Constraint12

Participation Constraint'

Constraint5
Constraint6
Constraint7
Constraint8
Constraint9
Constraint10
Constraint11

Co-Occurence Constraint'

Constraint1

Object Class Cardinality Constraint'

Constraint12

Object Class Name names Object Class

A | A |
B | B |
C | C |

Relationship Set Name names Relationship Set'

A r_1 B | A | r_1 | B |
B r_2 C | B | r_2 | C |
B C A | B | C | A |

Object Class has Connection to Relationship Set'

A	Connection1	A	r_1	B
B	Connection2	A	r_1	B
B	Connection3	B	r_2	C
C	Connection4	B	r_2	C
B	Connection5	B	C	A
C	Connection6	B	C	A
A	Connection7	B	C	A

Relationship Set Name includes Object Class Name'

A r_1 B A
A r_1 B B
B r_2 C B
B r_2 C C
B C A B
B C A C
B C A A

Connection has Participation Constraint'

Connection1   Constraint5
Connection2   Constraint6
Connection3   Constraint7
Connection4   Constraint10
Connection5   Constraint9
Connection6   Constraint11
Connection7   Constraint8

Relationship Set has Co-Occurence Constraint'

| B | C | A | Constraint1

Object Class has Object Class Cardinality Constraint'

| C | Constraint12

Figure A.20   A mathematical model instance for the OSA meta-model.

General Constraint'	General Constraint Representation'	Left Set'	Right Set'
Constraint2	$2 \le a$	Left-Set1	Right-Set1
Constraint3	$a \le 3$		
Constraint4	$b = \lvert B \rvert$		

CO-Pair'	OCC-Pair'	PC-Pair'	CO-Min'	CO-Max'	OCC-Min'	OCC-Max'
CO-Pair1	OCC-Pair1	PC-Pair1	1	1	4	8
CO-Pair2		PC-Pair2	b	*		
		PC-Pair3				
		PC-Pair4				
		PC-Pair5				
		PC-Pair6				
		PC-Pair7				

PC-Min'	PC-Max'
0	*
a	b
2a	a+b

Constraint constrains Object Class'

Constraint2	B
Constraint3	B
Constraint4	A
Constraint4	B
Constraint5	A
Constraint6	B
Constraint7	B
Constraint8	A
Constraint9	B
Constraint10	C
Constraint11	C
Constraint12	C

Constraint constrains Relationship Set'

Constraint5	A	r_1	B
Constraint6	A	r_1	B
Constraint2	A	r_1	B
Constraint3	A	r_1	B
Constraint4	A	r_1	B
Constraint7	B	r_2	C
Constraint10	B	r_2	C
Constraint2	B	r_2	C
Constraint3	B	r_2	C
Constraint11	B	C	A
Constraint9	B	C	A
Constraint8	B	C	A
Constraint1	B	C	A
Constraint4	B	C	A

General Constraint has General Constraint Representation'

Constraint2	$2 \le a$
Constraint3	$a \le 3$
Constraint4	$b = \lvert B \rvert$

Figure A.20 (Cont'd.)   A mathematical model instance for the OSA meta-model.

Object Class Name is member of Left Set'

B   Left-Set1

Object Class Name is member of Right Set'

C   Right-Set1

CO-Min is sub-part of CO-Pair'

1   CO-Pair1
b   CO-Pair2

CO-Max is sub-part of CO-Pair'

1   CO-Pair1
*   CO-Pair2

Left Set is sub-part of Co-occurrence Constraint'

Left-Set1   Constraint1

Right Set is sub-part of Co-occurrence Constraint'

Right-Set1   Constraint1

CO-Pair is sub-part of Co-occurrence Constraint'

CO-Pair1   Constraint1
CO-Pair2   Constraint1

OCC-Min is sub-part of OCC-Pair'

4   OCC-Pair12

OCC-Max is sub-part of OCC-Pair'

8   OCC-Pair12

OCC-Pair is member of Object-Class Cardinality Constraint'

OCC-Pair1   Constraint12

PC-Min is sub-part of PC-Pair'

0   PC-Pair1
a   PC-Pair2
2a   PC-Pair3
0   PC-Pair4
0   PC-Pair5
0   PC-Pair6
0   PC-Pair7

PC-Max is sub-part of PC-Pair'

b   PC-Pair1
a+b   PC-Pair2
*   PC-Pair3
*   PC-Pair4
*   PC-Pair5
*   PC-Pair6
*   PC-Pair7

PC-Pair is member of Participation Constraint'

PC-Pair1   Constraint5
PC-Pair2   Constraint6
PC-Pair3   Constraint7
PC-Pair4   Constraint8
PC-Pair5   Constraint 9
PC-Pair6   Constraint10
PC-Pair7   Constraint11

}

Figure A.20 (Cont'd.)   A mathematical model instance for the OSA meta-model.

Appendix B

Some Remarks on Specification, Design, and Implementation

Object-oriented Systems Analysis (OSA) supports analysis activities. It helps analysts study and understand a system and allows them to document the results of their understanding. It does not, however, include specification, design, or implementation, which are the other activities involved in system development. Although OSA does not include these activities it does serve as the foundation for them. While no one can address all these topics in enough detail in a single book, this appendix briefly discusses each one of these activities and explains how OSA serves as their foundation.

B.1 OSS: Object-Oriented Systems Specification

OSS helps a specification specialist formulate a system-development contract for a client. OSS activities include (1) system-boundary definition, (2) rapid prototyping, (3) performance and functional requirements specification, (4) interface specification, and (5) contract writing.

OSA serves directly as the basis for system-boundary definition. We can use an independent high-level object classes to identify the system implementation boundary. To specify general system behavior, we may provide a state net for this high-level, system object class. To specify how the system interacts with its environment, we can declare boundary-crossing interactions using interaction diagrams.

OSA directly supports rapid prototyping. Since object classes and relationship sets have a formal interpretation (see Appendix A), it is possible to develop an interpreter for ORMs. Using a default interface, the interpreter would be able to do data insertions, deletions, and modifications. In this sense, rapid prototyping is immediate, given only an ORM. Since state nets and interaction diagrams are based on natural language statements, they cannot be machine interpreted. To make them machine interpretable, we would first have to formalize them.

We can specify performance and functional requirements in OSA diagrams as constraints and notes. Real-time requirements, for example, may be specified by using the existing real-time modeling constructs in OSA.

Given an ORM, it is possible to generate a first-cut input and output interface for human-computer interaction. These interfaces may then be customized using prototyping tools. For nonhuman interfaces, interaction characteristics may be described by ORMs, state nets, and interaction diagrams.

OSA, augmented by specification information, can serve as a knowledge base for contract writing. Since this knowledge base would contain much of the information required for a contract, it should be possible to develop tools to help generate contracts from an OSA model.

B.2 OSD: Object-Oriented Systems Design

OSD helps a systems designer organize analysis concepts into object modules with good cohesion and coupling properties. OSD activities supported by OSA include (1) system formalization, (2) lexicalization, (3) normalization, and (4) encapsulation.

When an analysis is complete, some of the information will have already been formalized. For design we should fully formalize the information contained in the OSA model targeted for implementation. For each ORM, we should write the general constraints in a formal language. For state nets, we should replace high-level states and transitions with lower and even lower level states and transitions until they become "executable" in our target design domain. For interaction diagrams, we should rewrite the interaction activities in our OSA model targeted for implementation as access, modify, insert, delete, and request activities. By pushing the OSA model down to this "executable" level, a formal model for OSD can be represented using the original modeling constructs of OSA.

Lexicalization involves two activities. We first provide a domain definition for each object class that can be represented lexically. Second, we eliminate as many non-lexical object classes as possible. We do this by replacing each non-lexical object class whose objects have a one-to-one correspondence with the objects of a lexical object class by the corresponding lexical object class. A good example is the *Order* object class from our running example being replaced by the *Order Number* object class. See [Embley 1989] for an explanation of how this may be done.

Normalization for OSD is related to normalization techniques for relational databases. We remove redundancy by eliminating, reducing, or reorganizing relationship sets in ORMs. See [Embley 1989] for an explanation of how this may be done. We also remove redundancy from state nets and interaction diagrams by eliminating or reorganizing states, transitions, and interaction links.

Encapsulation produces object modules, which are the basic building blocks of OSD. To form object modules, we encapsulate declarative, behavioral, and service information such that each object module is maximally cohesive and minimally coupled with other object modules. For an explanation of how to determine module cohesion and coupling characteristics, see [Embley 1988].

An object module represents a logical abstraction that hides implementation details within it, defines all possible states for instances, and defines services that other object modules can use. Often, high-level object classes become a foundation for object module definition since high-level classes are used to form a similar abstraction of objects in OSA.

B.3 OSI: Object-Oriented Systems Implementation

OSI helps an implementor code and optimize a design for a particular hardware and software environment. The activities involved include (1) data and process distribution, (2) software reuse, (3) custom code development, (4) optimization, and (5) fine tuning.

For the most part, OSA supports implementation only indirectly by providing the basis for specification and design. System specification and design provide the basis for implementation. Once decisions about hardware and implementation software are made, we can map design and specification information into code. Ideally, some of the code would come from reuse libraries. Optimization and fine tuning make the system acceptably executable.

Usually, it will be easiest to map design and specification information based on OSA into a relational or object-oriented database system or into an object-oriented programming environment. Designs and specifications based on OSA, however, may also be mapped into a traditional programming environment.

As we conclude, we single out software reuse for a few more comments because OSA directly supports many software reuse activities, whereas it only indirectly supports the other implementation activities. We can use OSA to assist in software reuse in four ways. First, OSA provides directly for inheritance, including multiple inheritance. Inheritance hierarchies can be used to provide code for specializations as is commonly done in object-oriented programming environments. Second, we gain the greatest reuse leverage by reusing as early as possible in the development process. In addition to code reuse, we should also be able to take advantage of reuse opportunities during design and specification and even during analysis. Third, we can use OSA for domain analysis which opens the way for additional reuse opportunities [Prieto-Diaz 1985]. Fourth, we can use OSA to model a knowledge structure for software reuse. Having a knowledge structure simplifies reuse by making it easier to locate reuse components [Embley 1987].

Appendix C

References

ABRIAL, J. R. 1974. "Data semantics." In *Data Base Management*, ed. K. L. Koffeman. North-Holland, Amsterdam, The Netherlands, pp. 1–60.

ALFORD, M. W. 1977. "A requirements engineering methodology for real-time processing requirements," *IEEE Transactions on Software Engineering*, SE-3(1):60–69.

BALZER, R. M., GOLDMAN, N. M., and WILE, DAVID S. 1982. "Operational specification as the basis for rapid prototyping," *ACM Software Engineering Notes*. 7(5):3–16.

BATINI, C., LENZERINI, M., and NAVATHE, S. B. 1986. "A comparative analysis of methodologies for database schema integration," *ACM Computing Surveys*, 18(4):323–364.

CHEN, P. P. 1976. "The entity-relationship model—toward a unified view of data," *ACM Transactions on Database Systems*, 1(1):9–36.

CHVALOVSKY, V. 1983. "Decision tables," *Software-Practice and Experience*, 13(1):423–429.

COAD, P. and YOURDON, E. 1991. *OOA: Object-Oriented Analysis (2nd Ed.)*, Prentice Hall, Englewood Cliffs, NJ.

COAD, P. and YOURDON, E. 1990. *OOA: Object-Oriented Analysis*, Prentice Hall, Englewood Cliffs, NJ.

CODD, E. F. 1972. "Further normalization of the database relational model," in *Data Base Systems*, ed. R. Rustin, pp. 33–64, Prentice Hall, Englewood Cliffs, NJ.

CODD, E. F. 1970. "A relational model for large shared data banks," *Communications of the ACM*.

COOLAHAN, J. E. and ROUSSOPOULOS, N. 1983. "Timing requirements for time-driven systems using augments petri-nets," *IEEE Transactions on Software Engineering*, SE-9(5):606–616.

CZEJDO, B., ELMASRI, R., RUSINKIEWICZ, M., and EMBLEY, D. W. 1990. "A graphical data manipulation language for an extended entity-relationship model," *Computer*, 23(3):26–36.

CZEJDO, B. and EMBLEY, D. W. 1991. "View specification and manipulation for a semantic data model," *Information Systems*, 16(6):(in press).

DAVIDOFF, M. 1990. *The Satellite Experimenter's Handbook*, The American Radio Relay League.

DAVIS, A. M. 1990. *Software Requirements: Analysis and Specification*, Prentice Hall, Englewood Cliffs, NJ.

DAVIS, A. M. et al. 1979. "RLP: An automated tool for the automatic processing of requirements," *IEEE COMPSAC '79*, pp. 188–194, IEEE Computer Society Press, Washington DC.

DAVIS, C. G. and VICK, C. R. 1977. "The software development system," *IEEE Transactions of Software Engineering*, SE-3(1):69–84.

deCHAMPEAUX, D. 1991. "Object-oriented analysis and top-down software development," *Fifth European Conference on Object-Oriented Programming*, Centre Universitaire d'Informatique of the University of Geneva, Geneva, Switzerland.

DeMARCO, T. 1979. *Structured Analysis and System Specification*, Prentice Hall, Englewood Cliffs, NJ.

DYER, M. E. et al. 1977. *REVS Users Manual. SREP Final Report.*

ELMASRI, R. and NAVATHE, S. 1989. *Fundamentals of Database Systems*, Benjamin/Cummings, Menlo Park, CA.

ELMASRI, R., HEVNER, A., and WEELDREYER, J. 1985. "The category concept: an extension to the entity-relationship model," *Data & Knowledge Engineering*, 1(1):75–116.

EMBLEY, D. W., KURTZ, B. D. and WOODFIELD, S. N. 1991. *Solutions for Exercises in Object-Oriented Systems Analysis: A Model-Driven Approach*, Prentice Hall, Englewood Cliffs, NJ.

EMBLEY, D. W., LIDDLE, S. W. and WOODFIELD, S. N. 1991. "Attributes deemed detrimental in semantic data models," Technical Report BYU-CS-91-2, Department of Computer Science, Brigham Young University, Provo, UT.

EMBLEY, D. W., LIDDLE, S. W. and WOODFIELD, S. N. 1991. "Cardinality constraints in semantic data models," Technical Report BYU-CS-91-4, Department of Computer Science, Brigham Young University, Provo, UT.

EMBLEY, D. W. and LING, T. W. 1989. "Synergistic database design with an extended entity-relationship model," *Proceedings of the Eighth International Conference on Entity-Relationship Approach*, pp. 118–135, Toronto, Canada, 18–20 October 1989.

EMBLEY, D. W. and WOODFIELD, S. N. 1988. "Assessing the quality of abstract data types written in Ada," *Proceedings of the Tenth International Conference on Software Engineering*, pp. 144–153, Raffles City, Singapore, April 1988.

EMBLEY, D. W. and WOODFIELD, S. N. 1987. "A knowledge structure for reusing abstract data types," *Proceedings of the Ninth International Conference on Software Engineering*, pp. 360–368, Monterey, CA. March/April 1987.

GANE, C. and SARSON, T. 1979. *Structured Systems Analysis: Tools and Techniques*, Prentice Hall, Englewood Cliffs, NJ.

GOLDSTEIN, R. C. and STOREY, V. C. 1989. "Some findings on the intuitiveness of entity-relationship constructs," *Proceedings of the Eighth International Conference on Entity-Relationship Approach*, pp. 6–20, Toronto, Canada, 18–20 October 1989.

GRIES, D. 1985. *The Science of Programming*, Springer-Verlag, New York, NY.

HAREL, D. 1988. "On visual formalisms," *Communications of the ACM*, 31(5):514–530.

HATLEY, D. and PRIBHAI, I. 1989. *Strategies for Real-Time System Specification*, Dorset House.

Ho, D. 1990. "Object-oriented systems analysis: an update," *Proceedings of the 1990 Software Engineering Productivity Conference,* pp. 347–358, San Jose, CA, 21–23 August 1990.

Hull, R. and King, R. 1987. "Semantic database modeling: survey, applications, and research issues," *ACM Computing Surveys,* 19(3):201–260.

Jackson, M. A. 1983. *System Development,* Prentice Hall International, Englewood Cliffs, NJ.

Kung, C. 1990. "Object subclass hierarchy in SQL: a simple approach," *Communications of the ACM,* 33(7):117–125.

Kung, C. H. 1989. "Conceptual modeling in the context of software development," *IEEE Transactions on Software Engineering,* SE-15(10):1176–1187.

Kurtz, B. D. 1988. OSA: An object-directed methodology for systems analysis and specification. Master's Thesis, Department of Computer Science, Provo, Utah: Brigham Young University.

Kurtz, B. D., Ho, D. and Wall, T. A. 1989. An object-oriented methodology for systems analysis and specification. *Hewlett-Packard Journal,* 40(2):86–90.

Kurtz, B. D., Ho, D. and Parry, T. A. 1989. An object-oriented methodology for systems anlysis and specification. *Proceedings of the Hewlett-Packard Second European Software Engineering Productivity Conference,* pp. 269–277, Boebingen, Federal Republic of Germany, May 22–24, 1989.

Kurtz, B. D., Woodfield, S. N., and Embley, D. W. 1990. Object-oriented systems analysis and specification: A model-driven approach. *Proceedings of Compcon Spring 90,* pp. 328–332, San Francisco, Calif. February 27–March 1, 1990.

Ling, T. W. 1985. "An analysis of multivalued and join dependencies based on the entity-relationship approach," *Data & Knowledge Engineering,* 1:253–271.

Mark, L. 1983. "What is the binary relationship approach?" in *Entity-Relationship Approach to Software Engineering,* ed. R. T. Yeh, pp. 205–220, North-Holland.

McMenamin, S. M. and Palmer, J. F. 1984. *Essential Systems Analysis,* Yourdon Press, Englewood Cliffs, NJ.

Meyer, B. 1988. *Object-Oriented Software Construction,* Prentice Hall International, Cambridge, England.

Moret, B. 1982. "Decision Trees and Diagrams," *ACM Computing Surveys,* 14(4):593–623.

Orr, K. T. 1981. *Structured Requirements Definition,* Ken Orr & Associates, Inc., Topeka, KA.

Peckham, J. and Maryanski, F. 1983. "Semantic data models," *ACM Computing Surveys,* 20(3):153–189.

Peterson, J. L. 1981. *Petri Net Theory and the Modeling of Systems,* Prentice-Hall Inc., Englewood Cliffs, NJ.

Prieto-Diaz, R. 1985. "A software classification scheme," Ph.D. Dissertation, Department of Information and Computer Science, University of California—Irvine, Irvine, CA.

Ross, D. T. and Schoman Jr., K. E. 1977. "Structured analysis for requirement definition," *IEEE Transactions on Software Engineering,* vol. SE-3(1).

Rumbaugh, J., Blaha, M., Premerlani, W., Eddy, F. and Lorensen, W. 1991. *Object-Oriented Modeling and Design,* Prentice Hall, Englewood Cliffs, NJ.

Shlaer, S. and Mellor, S. J. 1988. *Object-Oriented Systems Analysis: Modeling the World in Data,* Yourdon Press, Englewood Cliffs, NJ.

Smith, J. M. and Smith, D. C. P. 1977. "Database abstractions: aggregation and generalization," *ACM Transactions on Data Base System,* 2(2):105–133.

TAYLOR, B. J. 1980. "Introducing real-time constraints in requirements and high-level design of operating systems," *National Telecommunications Conference,* pp. 18.5.1–18.5.5, IEEE Computer Society Press, Washington DC.

TEICHROEW, D. and HERSHEY, III, E. A. 1977. "PSL/PSA: A computer-aided technique for structured documentation and analysis of information processing systems," *IEEE Transactions on Software Engineering,* SE-3(1):41–48.

TEOREY, T. J., WEI, G., BOLTON, D. L. and KOENIG, J. A. 1989. "ER model clustering as an aid for user communication and documentation in database design," *Communications of the ACM,* 32(8):975–987.

TEOREY, T. J., YANG, D. and FRY, J. P. 1986. "A logical design methodology for relational databases using the extended entity-relationship model," *ACM Computing Surveys,* 18(2):197–222.

WANG, Y. 1988. "A distributed specification model and its prototyping," *IEEE Transactions on Software Engineering,* SE-14(8):1090–1097.

WARD, P. T. and MELLOR, S. J. 1985. *Structured Development for Real-Time Systems (Volume 1–3),* Yourdon Press.

WARNIER, J.-D. 1981. *Logical Construction of Programs,* Van Nostrand-Reinhold.

WARNIER, J.-D. 1974. *Logical Construction of Programs,* Van Nostrand-Reinhold.

WASSERMAN, A. I., PIRCHER, P. A., SHEWMAKE, D. T. and KERSTEN, M. L. 1986. "Developing interactive information systems with the user software engineering methodology," *IEEE Transactions on Software Engineering,* SE-12(2):326–345.

WASSERMAN, A. I. 1985. "Extending state transition diagrams for the specification of human-computer interaction," *IEEE Transactions on Software Engineering,* 11(8):669–713.

YOURDON, E. 1989. *Modern Structured Analysis,* Yourdon Press, Englewood Cliffs, NJ.

ZAVE, P. and SCHELL, W. 1986. "Salient features of an executable specification language and its environment," *IEEE Transactions on Software Engineering,* SE-12(2):312–325.

ZAVE, P. 1982. "An operational approach to requirements specification for embedded systems," *IEEE Transactions on Software Engineering,* SE-8(3):250–269.

Index

@ — Event representation, 10, 64
| — Multiple name separator, 25

A

Action, 10, 62
  destroying action, 78
  generalization-specialization, 90–91
  implies access interaction, 179
  in transition views, 118
  initializing action, 77
  interruptible, 62
  non-interruptible, 62
  representation, 64
  real-time constraint on, 85–86
  shorthand representation, 79
  timing, 64–65

Activity represented by state, 60–62
Aggregate. (*See* Superpart)
Aggregate class. (*See* Superpart class)
Aggregation, 38, 44–46
  aggregate class, 44
  component class, 44
  and high-level object class, 135
  and high-level states, 154
  multiple decompositions, 45–46
  participation constraints, 44
  and relational object class, 135
  represented by black triangle, 44
  and state nets, 154
  subpart class, 44
  superpart class, 44
Amateur radio satellite example, 236–52
  clocks, 239–40
  special notes, 240

I